Multiple Stable States in Natural Ecosystems

Multiple Stable States in Natural Ecosystems

PETER PETRAITIS
Professor of Biology, University of Pennsylvania

Great Clarendon Street, Oxford, OX2 6DP,
United Kingdom

Oxford University Press is a department of the University of Oxford.
It furthers the University's objective of excellence in research, scholarship,
and education by publishing worldwide. Oxford is a registered trade mark of
Oxford University Press in the UK and in certain other countries

First Edition published in 2013
Impression: 1

British Library Cataloguing in Publication Data
Data available

ISBN 978–0–19–956934–2

Printed and bound by
CPI Group (UK) Ltd, Croydon, CR0 4YY

For Carol, who has been with me for each step,
"no me sostiene el pan, el alba me desquicia
busco el sonido líquido de tus pies en el día"
(Sonnet XI, Pablo Neruda)

Preface

The notion of multiple or alternative community states has captured the imagination of ecologists for over 50 years, and this book is only a snapshot of what has been done by so many researchers. Multiple stable states has been discussed and tested by researchers across many subdisciplines in ecology and unfortunately without much exchange of ideas across the borders. The idea of multiple stable states in ecosystems has been discovered with near independence or invented anew by researchers studying lakes, coral reefs, and semi-arid rangelands. There has been some cross-fertilization, but not as much as we might hope.

A casual glance at the citations of five of the most influential and seminal papers as of August 2012 easily makes lack of cross-fertilization clear (e.g. Lewontin 1969, May 1977, Noy-Meir 1975, Scheffer et al. 2001, Westoby et al. 1989). The ends of the spectrum are Lewontin's paper, which has been cited about 150 times, and Scheffer et al.'s paper, which has been cited over 1450 times. I suspect the frequency of citation is a reflection of the ease of access and the rise of reference databases and software. Lewontin's paper was published in a symposium volume, which is still relatively difficult to obtain. In between are Noy-Meir's paper with just over 400 citations, May's paper with almost 550, and Westoby et al.'s paper with about 750. It is interesting that the rank correlation between the number of citations and the date of publication is perfectly negative (i.e. −1.0). Far more troubling is the lack of overlap in citations. May's paper introduced the idea of multiple stable states to a broader audience, and Westoby et al.'s paper lays the groundwork for state-and-transition models and is the touchstone for nearly all models of multiple stable states of grazing in grasslands and rangelands. May's and Scheffer et al.'s papers are most often cited by researchers not working in rangelands; researchers of rangeland systems rarely cite May or Scheffer et al. There are only about 40 papers in which both Westoby et al. and Scheffer et al. are cited and another 40 papers that cite both May and Westoby et al. Out of roughly 2,300 unique citations among these three papers, only ten or about 0.4% cite all three.

As a result, I have had to make a number of tough decisions given the overwhelming span of papers about multiple stable states, and for the most part, I have attempted to cite only the earliest and more relevant literature. I have, indeed, left many citations out for various reasons.[1] As Oscar Wilde said, "I am not young enough to know everything."

This book has had a long gestation, and I am indebted to many people and organizations that have supported me during the process of writing. The beginnings of the book

[1] References can be overlooked for a variety of reasons, both pro and con. Failure to cite someone may reflect an unwillingness to cite poor, but "important" work as well as overlooking of important but obscure research. The h-index may tell us more about a Matthew effect than anything else.

took place during my sabbatical in Chile, and I am especially grateful for the support from the Fulbright Foundation and Pontificia Universidad Católica de Chile while on sabbatical. Much of the earlier work on the book could not have been accomplished without the warm and generous support of my friend and host, Juan Carlos Castilla in Chile. I also would like to thank the US National Science Foundation for its support of my research on multiple stable states. I am grateful for the patience of Ian Sherman, Lucy Nash, Helen Eaton, and the rest of the staff of Oxford University Press for seeing me through to the end. I also thank Mike Angilletta for suggesting that Oxford University Press might be a good home for this book. Parts of several sections in Chapters 2, 4, 8, and 9 have been re-worked from parts of published papers (Petraitis and Dudgeon 2004, Petraitis and Hoffman 2010, Petraitis et al. 2009).

There are many people who have pushed me along on this path. My mentors played an important role in my development as an ecologist and in how I think about ecology, and I am indebted to Tom Ebert, Jeff Levinton, and Fred Grassle. My early speculations about multiple stable states were fueled by conversations with Roger Latham and Peter Fairweather in the early 1990s. Roger and Peter have continued to provide insightful comments. Almost all of my research on multiple stable states has been done in collaboration with Steve Dudgeon, and I am especially grateful for his constant stream of encouragement, comments, and help over the last 15 years. Jon Fisher (Memorial University, Newfoundland), B. Boldgiv (National University of Mongolia), Mike Sears (Clemson University), and Mike Russell (Villanova University) have also provided many insights as have the students and faculty of the ecology and evolution group in the Department of Biology of the University of Pennsylvania. Much of my work in the field, which allowed me to think more deeply about multiple stable states, could not have been done without help from Erika Rhile, B. Boldgiv, Jon Fisher, Steve Dudgeon, Catharine Hoffman, Lisa Methratta, Nick Vidargas, Annalise Paaby, and the many students from Cheverus High School who were brought to Swan's Island by their teacher, Erika Rhile.

The writing of the book occurred in short bursts over many years, and I must thank some of my friends who provided me with places to write. This includes not only many nice desks at various universities in the USA, Chile, Australia, and Mongolia but also the occasional kitchen table in the homes of friends. For providing me this collection of quiet places, I would like to thank Peter Fairweather and Gillian Naiper (Adelaide, Australia), Tony Underwood and Gee Chapman (Spain), Jim and Wendy Hunt (San Diego, California), Bob Steneck and Rick Wahle (Darling Center, University of Maine), Gary and Mimi Rainford (Swan's Island, Maine), B. Boldgiv (National University of Mongolia), Juan Carlos Castilla (Pontificia Universidad Católica de Chile), Sergio Navarrete (Estación Costera de Investigaciones Marinas in Las Cruces, Chile), Ricardo Guiñez (Universidad de Antofagasta, Chile), Martin Thiel and Carlos Gaymer (Universidad Católica del Norte, Coquimbo, Chile), and Alvaro Palma (Quintay, Chile).

I could not have written this book without the love, support, and encouragement of my family. Carol, Dan, and Rob have always kept me grounded and reminded me that doing an experiment in ecology is but a small slice of a world that includes bigger things such as making music, creating art, and working for human rights. Thank you for keeping me focused on the truly important things in life.

Peter Petraitis
27 August 2012
Swan's Island, Maine

Contents

1 Introduction

One of the most interesting and vexing problems in ecology is how natural ecosystems can appear to be so persistent and yet seem so susceptible to catastrophic change. This duality of persistence and susceptibility is observed in a wide variety of ecosystems, from coral reefs that are suddenly replaced with seaweeds to semi-arid grasslands that are shifted to shrub forests. Often the shifts are not only quite dramatic and occur without warning but also are very difficult to reverse. How can these communities appear to be so persistent for long periods, but then small shifts in environmental conditions cause an unexpected tipping of the system?

Ecologists are especially interested in these sudden shifts for both conceptual and practical reasons. On the conceptual side, these characteristics are often seen as indicators of an ecosystem with multiple stable states (e.g. de Young et al. 2004) while on the practical side, the unpredictable nature of the shifts and the inability to reverse changes can cause serious management problems (e.g. Bestelmeyer 2006, Groffman et al. 2006, Huggett 2005, Suding and Hobbs 2009, Thrush et al. 2009).

Both these characteristics—sudden and dramatic shifts in species composition and difficulty in reversing the shift—are considered hallmarks of multiple stable states, and so the theory of multiple stable states is often offered as an explanation for these sudden and unexpected shifts in the species composition of ecosystems. Again and again, natural communities that have undergone abrupt and difficult to reverse changes are put forward as examples of ecosystems with multiple stable states. Yet there are few clear guidelines about what constitutes good evidence for multiple stable states. How do we know if a sudden shift in species composition is evidence for multiple stable states, and more importantly, how do we define terms such as "sudden," "abrupt," "persistent," and "irreversible," which are so often used to discuss multiple stable states in natural communities?

1.1 Two examples of sudden irreversible shifts

Many of these abrupt and nearly irreversible changes have been due to the activities of humans. Yet while humans have been transforming the landscape and causing dramatic shifts in natural communities for hundreds, if not thousands, of years, in many cases it is not known if these shifts represent multiple stable states. The clearing of southern Australia and the demise of the passenger pigeon in North America provide compelling examples in which it is clear that the system has undergone irreversible changes that are likely to have some of the hallmarks of systems with multiple states.

Unlike many of the better known examples, the transformation of southern Australia is well documented. When Europeans began settling southern Australia, in the 1860s, they were confronted with vast expanses of wet eucalypt-dominated jungle that extended eastward across the Gippsland Plains and into the Strzelecki Ranges. Today this region, which is near Melbourne, is mostly a mixture of pasture, eucalypt savanna, and meadows of bracken fern and blackberries, which are nonnative species. The original forest was home to the world's largest flowering plant, *Eucalyptus regnans*, the Australian mountain ash, which can reach heights in excess of 100 m. Tarra-Bulga National Park contains one of the few remaining patches of mountain ash forest and gives one a sense of what settlers faced throughout the entire Gippsland Plains. The remaining trees of mountain ash in Tarra-Bulga tower over an understory of other large species such as blue gum (*Eucalyptus globulus*), and myrtle beech (*Nothofagus cunninghamii*). Tree ferns reach 20 m in height, and the wet ground is blanketed with smaller ferns and is home to the giant Gippsland earthworm, *Megascolides australis*, which can reach 3 m in length.

When settlers began to clear the forest for pasture, they did so with remarkable efficiency. On the plains, the trunk of each large tree was cut and girdled, cutting off the flow of sap from the roots to the crown, and low-hanging branches were cut from the trunk. Long, narrow planks, called springboards, were pounded into notches in the trunk and provided a series of stairs so that settlers could climb and cut the higher limbs. During the dry period of summer, the settlers organized community-wide burns to clear out the cut limbs and the dead trunks, and the resulting fires would often burn out of control for weeks.

The steep slopes were cleared with even greater ease: "A spectacular and economic felling method was quickly developed in the hilly country. Work would start from an already cleared pack track at the top of a ridge. Trees all the way down to the bottom of the gully would be 'nicked' on the up hill side. Then a few big trees high up the slope would be felled with care and accuracy to start an avalanche of spars (trunks), which bounced and crashed and roared down hill." (Daley 1960, p. 97).

Within a very short time, the landscape was transformed—cleared of forest and converted into pasture. The Gippsland Plains and the western areas of the Strzelecki Ranges were cleared by the 1890s. Many areas, particularly in the eastern Strzelecki Ranges, proved to be too steep for pasture and were abandoned only to be taken over by bracken fern and blackberries.

The loss of the forest has completely altered the evapotranspiration regime of the region. The cool wet forest with its boggy soil is now open dry land. Gippsland's giant earthworm is rare although honored in the annual worm festival in Korumburra, Victoria. Bracken fern and blackberries are a constant problem, even in Tarra-Bulga National Park. The mountain ash has not recovered and is unlikely to do so. Unlike many eucalypt species, mountain ash can be killed by severe fires and does not have the capacity to regenerate after fire. The species depends on seed for regeneration, and the near perfect clearing of the forest has not only cut off the input of seeds but also created an environment in which germination is difficult. The shift from forest to pasture, bracken, and blackberries has been swift and has proven difficult to reverse.

The difficulty in reversing this trend is likely to be a combination of several factors (Lindenmayer et al. 2011). Specifically addressing mountain ash, Lindenmayer et al. suggest that landscape-level processes are important, and include climatic effects that have altered life-history patterns. Interestingly, they coin a new phrase for this—landscape traps—and assert that these are fundamentally different from multiple stable states. In fact, their conceptual model looks strikingly similar to models of multiple stable states that have been proposed for semi-arid rangelands and coral reefs.

The disappearance of the passenger pigeon (*Ectopistes migratorius*) was just as dramatic. In a span of 100 years, the passenger pigeon was driven from being the most abundant and common bird in North America to extinction. At the time of the arrival of Europeans in North America, the population of passenger pigeons was estimated to be between 3 and 5 billion (10^9), possibly accounting for 25 to 40% of all the birds in North America (Schorger 1955). Passenger pigeons ranged as far west as Texas, south to northern Florida, and into southern Canada. Nesting grounds were confined to southern New England and the US states surrounding the Great Lakes. The appellation "passenger" refers to the bird's migratory behavior and likely from the French adjective *passager* which means passing.

Audubon's description of the massive numbers of passenger pigeons around 1830 is almost unbelievable (Sanders 1986). Audubon, who is well known for his exquisite paintings of North American birds, described traveling by horseback for 3 days and over a distance of 90 km and having the sky filled with birds the entire time. The birds were so numerous that the sky was darkened as during an eclipse and droppings fell like snow.

Passenger pigeons formed large communal roosts at night and for breeding. Nesting pairs produced one egg, and both parents produced "milk" which was regurgitated from the crop (Schorger 1955). A nesting roost in Michigan in 1879 covered approximately 26 000 hectares (Halliday 1980). Pehr Kalm, who was a student of Linnaeus and was in North America in the 1740s, wrote that it was dangerous to walk through a forest during nesting (Kalm and Gronberger 1911). The weight of roosting birds could be so great that tree limbs the "size of a man's thigh" could break off and whole trees were often uprooted. Droppings could be 0.3–0.4 m thick on the ground.

Passenger pigeons ate the fruits and nuts of at least 29 species (Webb 1986). Their diet varied with season, and during the spring could include insects. Schorger (1955) mentions that birds would dis-engorge food if they found a more desirable item, and Webb (1986) suggests the range expansion of trees with large seeds after the retreat of the Wisconsin ice sheet 10 000 to 15 000 years ago was facilitated by passenger pigeons.

The impact of their feeding must have been enormous. Passenger pigeons weighed between 340 and 400 g and based on estimated field metabolic rates, they would have required between 550 and 620 kJ per day per bird (Nagy 2005). Large flocks were capable of stripping oak forests of nearly all acorns during mast years, and it should not be surprising that early settlers feared the impact of passenger pigeons on their crops (Schorger 1955).

Given the ease of capturing and killing passenger pigeons, they were exploited as food. Birds were shot—a single discharge of a shotgun was capable of killing up to 100 birds. Audubon said that it was often not possible to hear the discharge over the cooing of the birds. Trapping was done using captive birds, which were tethered to platforms and attracted other birds. These tethered birds were known as "stool pigeons." Given the ease of capture, the industry included wholesale dealers who would provide information about roosting sites and helped spur of the construction of railroads near roosting sites to lower the costs of shipping.

By the early 1890s, passenger pigeons were nearly gone. Conservation efforts failed to slow the decline. Schorger (1955) provides a listing of reported sightings of a few birds here and there during final decade, and stated in 1900 the last bird in the wild was killed in Ohio. A few birds survived in the Cincinnati Zoo but never bred successively, and the last bird, named Martha after Martha Washington, died in 1914.

The decline of the mountain ash and loss of the passenger pigeon contain features that are often associated with multiple stable states. For the mountain ash it is clear that there are two states—forests with the tree and more savanna-like systems without it. The alternative state for passenger pigeons is a "state" in the sense the bird no longer exists. In both cases, there appears to have been a point of no return at which decline in the system cannot be reversed. For animals, this is known as the Allee effect. Allee (1931) suggested that there can be a beneficial result of social interactions—for example group behavior that lowers the per-capita death rate due to predation. This effect, however, can be reduced if population levels fall too low, and thus is one of the common explanations for the decline of passenger pigeons. Communal roosts became too small to support successful breeding. While Allee was primarily interested in social interactions, any plausible ecological process that gives rise to a nonlinear function for either births or deaths has the potential for creating a system with multiple equilibrium points. The case of mountain ash is also an Allee effect in that recruitment of seedlings fails at low densities of adult trees.

Tied to the point of no return is the phenomenon of hysteresis, which is the existence of sudden and difficult to reverse shifts. The term was coined in 1890 by Sir James Alfred Ewing, from the ancient Greek word for deficiency or lagging behind, during his studies of the magnetization of metals induced by electric currents. Ewing noted that when the he reversed the direction of an electric current the reversal of the polarity of induced magnetic field in metals was not instantaneous and lagged behind the change in current direction. This lag is due to a nonlinear response in the system—Ewing hypothesized that individual metal particles retained a memory of the electric field and resisted reversals in current. Formal development of the mathematics of hysteresis in nonlinear systems was carried out in the 1970s by a group of Russian mathematicians led by the Ukrainian Mark Krasnosel'skii (Krasnosel'skii and Pokrovskii 1989). Currently the term hysteresis is more broadly used for any system in which the response to an object, a material, or a system lags behind the force applied to it. However, the underlying mathematics of all examples contains some sort of nonlinearity.

1.2 Making the connections between observations and theory

Are the changes from mountain ash forests to pasture and the extinction of the passenger pigeon reflections of the fact that these systems contain multiple stable states? And more importantly, how would we know? These questions require us to examine what we can learn from the theory of multiple stable states and how theory can be used to guide us in developing critical experimental tests and in identifying key observations. We must also come to terms on how we translate the terminology of models—parameters, state variables, and stability—into useable concepts for experimental ecology.

Nearly 40 years ago, theorists provided plausible mathematical models to explain how different communities could arise in the same environment (Lewontin 1969, May 1977, Noy-Meir 1975). Lewontin (1969) asked, "Can there be more than one stable community composition in a given habitat?" His answer was yes, and he along with others (Knowlton 1992, May 1977) went on to suggest that small environmental perturbations can produce large, discontinuous, and irreversible changes in natural communities. The switch between different communities is often sudden and occurs at a critical environmental threshold with a rapid transition in species composition.

Models of multiple stable states are easy to construct. The simplest and most often cited example in ecology is the unstable coexistence of two competing species in the Lotka–Volterra model of competition. In this case, each species, once established, can resist the invasion by the other, and these alternative conditions represent the two stable points of the system. Past history matters, and the first species that arrives and becomes established will be able to prevent any attempts by the other to invade. If both species arrive at the same time, then the relative abundance of the species upon arrival and their characteristics (i.e. the *r*s and *K*s and competition coefficients) will determine which species wins. This situation is often described as two bowls or valleys separated by a ridge. The two stable equilibrium points are at the bottom of the bowls and on the ridge there is a third equilibrium point that is unstable. This pattern was described by Lotka (1956, p. 147), who wrote, "it is clear that two pits of [stable equilibrium points] . . . cannot occur without some other type of singular point between them, just as it is physically impossible for two mountains to rise from a landscape without some kind of a valley between", although he warned this analogy was "purely qualitative." Almost as an aside, Lotka mentions that the reader can find a more complete discussion of the mathematics involved and cites references as early as 1891. While Lotka seems to have mixed his metaphor of pits with mountains, ecologists now visualize multiple stable states as valleys separated by a ridge.

The possibility of more complex models for multiple stable states in ecosystems first came to the attention of ecologists with the publication of Lewontin's article "The meaning of stability" in 1969, but it was May's 1977 review article in *Nature* that brought the notion of multiple stable states to the forefront of ecology. May developed a series of models that showed how ecological systems could undergo abrupt and dramatic changes, and paraphrasing a 1660 saying by Pascal, May suggested these changes were as if the "hinge of history turn[ed] on the length of Cleopatra's nose...." May used dynamical models to explore catastrophic shifts in budworm outbreaks on spruce, crashes in harvesting in commercial fisheries, and transmission of malaria and schistosomiasis, and his demonstration that relatively simple models could give rise to complex and unexpected outcomes captured the imagination of ecologists. In contrast, Lewontin (1969) saw the possibility of catastrophic shifts as a serious problem for developing predictive theory in ecology and added, "[I]t must be hoped that such structural instabilities will prove to be the exception."

While the theory of these systems is well understood, whether multiple stable states actually exist in nature has remained a hotly debated subject, and the translation of theory into experimental tests has not been easy. The phrase "multiple stable states" has a very precise meaning in theory, but not in practice. Lewontin stressed that random events and historical accidents blur what stability, equilibrium, and habitat mean in nature. While stability is well defined by theorists, there is little agreement on how stability should be measured in nature (Grimm and Wissel 1997). The modifiers used to describe shifts in species composition add to the problem. Labeling shifts in species composition as "dramatic," "sudden," or "unexpected" is akin to Shakespeare asking, "Shall I compare thee to a summer's day?" (Sonnet 18). How big must a shift be to be called dramatic, or how fast to be sudden? Is it only the lack of information that causes us to be surprised by unexpected changes in species composition? Differences in terminology among subdisciplines also present serious problems. The idea that multiple stable states underlie the observation of sudden and nearly irreversible shifts in natural communities has been elaborated in several different areas of ecology, with each subdiscipline using different terms for the same phenomena, or worse, the same term for different phenomena. To add to the confusion, several conceptual ideas—most notably, state and transition models and regime shifts—have been broadened to include multiple stable states. Yet systems with regime shifts or models using states and transitions do not always have multiple stable states. Finally, while reviews of the literature tend to agree that the experimental evidence is conflicting, the same experiment may be cited as supporting opposing views (Connell and Sousa 1983, Didham et al. 2005, Peterson 1984, Scheffer et al. 2001, Schröder et al. 2005, Sousa and Connell 1985).

Part of the difficulty is how we make the links between theory, observation, and experiments. Bob May, in his seminal paper on multiple stable states (May 1977), even commented that "empirical observations remain largely anecdotal, and... theory remains metaphorical." I would suggest that even today May's observation is apropos for almost all areas of ecology, and I believe making the link between anecdotes and metaphors remains one of the biggest challenges facing ecologists. At best, ecological theory provides ideas about how to organize observations of natural phenomena and guides research by providing testable predictions. The utility of a particular theory in ecology depends not only on how well the theory "works" but also the assumptions we make in order to create the theory. The use of simplifying assumptions means that ecological theory will always be only a metaphor or a map for reality. Not all metaphors or maps are equally useful, and a trivial but compelling example is the fact that a subway map for New York City is not useful for driving in Manhattan.

If theory provides us with maps then the usefulness of a particular map depends on what the map is meant to represent, how well it does so, and an implicit agreement between the mapmaker and the map reader about the representation. A New York subway map is a beautifully drawn representation, but not all maps are as well drawn. Even in this day and age of Google maps, most of us have suffered from the fate of relying on a map hastily drawn by a well-intentioned friend with directions to a party. Our friend's sense of distance, important landmarks and even sense of direction may not match our own, and so we missed a critical turn and ended up arriving late.

Not all theoreticians are skilled mapmakers and not all experimentalists are adept at reading maps or using them to navigate. For example a misreading of the map for competition that was formulated by Lotka and Volterra led ecologists on a long detour about competitive exclusion. The mathematical model of two-species competition developed by Volterra and Lotka indicated that range of parameter values that allowed for stable coexistence of both species was small, and species with similar requirements could not coexist. The clearest and most precise statement of this finding is "Complete competitors cannot coexist" (Hardin 1960). However, many ecologists used Gause's reading of Lotka and Volterra's map. Gause showed how difficult it was to maintain coexistence in a series of experiments on competition between two species of *Paramecium*, and in his book *The struggle for existence*, which was published when he was 24, he stated that two species could not share the same niche (Gause 1934). This interpretation is often called Gause's principle of competitive exclusion, even though several people stated the principle before him, which Gause himself acknowledged. Today, Grinnell is often cited as the earliest reference (Hardin 1960) although Johnson clearly stated the same idea in 1910 (Gaffney 1975). The interpretation by Gause, Grinnell, and Johnson of the map provided by Lotka and Volterra led ecologists on a 50-year expedition to understand why so many species appear to coexist in systems that seem to have very few niches or types of resources. Yet the problem is two-fold: the map was poorly drawn and the reading was imprecise. By the 1980s, theoreticians had shown that the number of coexisting species at stability could far exceed the number of resource classes in closed systems (e.g. Armstrong and McGehee 1980) if two key assumptions of the Lotka–Volterra model were relaxed. First, allow population growth to be a curvilinear function of densities, and second, consider nonpoint equilibria. If these conditions are met, then multiple species can coexist on as few as one or two resources. Interestingly Hardin's more precise statement still holds.

The theory of multiple stable states and the evidence for multiple stable states in nature embodies many of the same conceptual issues that surrounded the early attempts to understand competitive exclusion. This is not a new problem in ecology. Moving from theory to experiment is difficult in many areas of ecology, and in community ecology perhaps more so than in other areas where theory is often more of a metaphor than a map for what we see in nature. There are also often differing viewpoints of what constitutes a valid experimental test of theory and how theories as maps and metaphors can enhance or impede experimental work. Our examination of multiple stable states provides a compelling case study of how we might make the links between theory and the experimental evidence because the theoretical predictions are very striking and unique even though the development of good experimental tests appears to be difficult.

1.3 Overview of the book

The other nine chapters of this book are an eclectic mix of what is well known to ecologists and a number of things that have been overlooked. The next two chapters address the question of what theory actually tells us about multiple stable states. We will explore dynamical models of one, two, and three species to introduce the ecological view of the theory of multiple stable states. Models are covered with the specific goal of identifying testable predictions about multiple stable states. The number of equations has been kept to the bare minimum needed to describe the theory fully, and this overview of theory should be transparent to undergraduates who have taken an advanced ecology course.

The discussion in Chapter 2 will suggest that the presence of multiple basins of attraction—those pairs of valleys separated by a ridge that were envisioned by Lotka—is the structural property that uniquely defines a system with multiple stable states. This notion gives rise to the suggestions of Peterson (1984) of what constitutes a strong test of multiple stable states.

If Chapter 2 is a summary of the "maps" that have been created for ecosystems for multiple stable states, then Chapter 3 is a discussion of how to use those maps. Chapter 3 develops the specific predictions and approaches arising from the theory of multiple stable states. There are two broad categories of approaches: experiments that show two or more unique species assemblages can occur in the same environment and the detection of hysteresis. Both approaches are plagued with the difficulties of translating metaphors into clear unambiguous tests.

The question of stability in natural ecosystems is probably the thorniest issue. Grimm and Wissel (1997) complied an astounding list of 163 definitions for stability in ecological systems and managed to group these definitions into six distinct classes of proxies for stability—constancy, resilience, persistence, resistance, elasticity, and domain of attraction. We have not made much progress in finding a good ecological proxy for stability since all six proxies can be found in Lotka (1956) and Lewontin (1969).

There are depressingly few examples of good experimental tests of multiple stable states. One review of the literature turned up only 35 studies that were judged to be good tests with only nine field studies finding evidence consistent with multiple stable states (Schröder et al. 2005). In Chapter 4 we will take a closer look at some of these studies and explore three of them in detail, one of which was cited by Noy-Meir (1975) as the earliest study of multiple stable states.

Chapters 5 and 6 reintroduce catastrophe theory to ecologists. Catastrophe theory was developed by the French topologist René Thom and became known to English speakers in the 1970s. Catastrophe theory is the theory of multiple stable states, but somehow it has been overlooked by ecologists. While Thom and his ideas were controversial, by the 1980s the use of catastrophe theory in mathematics, physics, engineering, and chemistry was well established and well within the mainstream (Arnol'd 1992, Gilmore 1981, Poston and Stewart 1978, Zeeman 1976). All of the hallmarks of multiple stable states, including the newer ones of critical slowing down, anomalous variances, and autocorrelation, which have been recently rediscovered by ecologists, were well known to mathematicians, physicists, and engineers by the 1980s. We will explore why ecologists have overlooked this rich literature and discuss several important details that ecologists continue to ignore.

Chapters 7 through 9 will touch upon other modeling approaches that have been used, several common misconceptions, and the use of spatial and temporal patterns as evidence of multiple stable states. Models can provide the most compelling examples of plausibility of multiple stable states and hysteresis, but they are often difficult to confirm. As for misconceptions, we will see that multiple stable states in ecosystems do not require positive-feedback loops, ecosystem engineers, environmental switches, and stressful abiotic conditions as have been suggested. Our overview will include a close look at state-and-transition models (Westoby 1980, Westoby et al. 1989). State-and-transition models are often linked to multiple stable states, but it is quite clear that the idea of states and transitions was developed as a method to manage systems not at equilibrium and not as a specific model of multiple stable states.

In Chapter 10 we will evaluate the strengths and weaknesses of the various approaches, discuss why so little progress seems to have been made, and provide suggestions of how to move forward. Some types of systems are clearly better than others for undertaking experimental studies of multiple stable states, and the characteristics of ecosystems that are good candidates will be discussed. These features include the presence of ecosystem engineers that structure different assemblages, variation in early successional events, and ecological processes that occur on manageable temporal and spatial scales. We will explore the challenges that remain, especially in relation to designing better experiments, increasing access to primary data, and developing more precise hypotheses.

2

What theory actually tells us about multiple stable states

The idea of multiple stable states in ecological communities has its roots in the modeling of ecological processes by ordinary differential equations, and so we will use this approach to examine the insights that theory has provided to ecology about the conditions required for multiple stable states, bifurcation folds, and hysteresis. Theory quickly outstrips what experimentalists can test, and thus our discussion of theory will be limited to one-, two-, and three-species models of differential equations. These are more than sufficient to illustrate the basic framework. There are a number of other ways in which ecologists have tackled the issue of multiple stable states, such as matrix models and state and transition models. Both will be covered in Chapter 7 in our discussion of other approaches to modeling that have been used to demonstrate the plausibility of multiple stable states in natural communities. Regardless of the modeling approach that is used, there is gap between theorists and experimentalists about what can be inferred, and a brief introduction to what both theorists and experimentalists normally mean by models, variables, parameters, and equilibrium is necessary.

2.1 Models, state variables, parameters, and equilibrium points

It is useful to begin with a discussion of models, parameters, and state variables. Many of the better known models in ecology are of the form

$$\frac{\mathrm{d}N}{\mathrm{d}t} = f(N) \tag{2.1}$$

where the rate of change is a function of density or some other ecosystem state such as biomass or resource abundance. Throughout we will present the models in terms of density, but our conclusions can be generalized to other sorts of ecosystem descriptors. In these kinds of models, the rate of change is a function of density, which is the state variable, and the function itself can involve either parameters that cannot be directly measured, such as r and K in the logistic model, or parameters such as births and deaths that can be measured directly. Alternatively, many ecological models are of the form

$$\frac{dN}{dt} = f(R) \tag{2.2}$$

in which the rate of change is a function of a resource or something else rather than the state variable of interest. These sorts of models are often called "mechanistic" because it is assumed that the underlying mechanism for conversion of resources into population growth is understood and captured in the function $f(R)$. In ecological systems, variations are commonly observed in state variables (e.g. densities, biomass, resource levels, etc.) and in parameters (e.g. r, K, births, deaths, per capita rate of prey capture, etc.).

Typically the relationships among different species are expressed as sets of differential equations, and the standard Lotka–Volterra model for competition between two species is a familiar example of a pair of linear differential equations where

$$f_1(N_1, N_2) = \frac{dN_1}{dt} = r_1 N_1 - \frac{r_1}{K_1} N_1^2 - \frac{\alpha r_1}{K_1} N_1 N_2 \tag{2.3}$$

$$f_2(N_1, N_2) = \frac{dN_2}{dt} = r_2 N_2 - \frac{r_2}{K_2} N_2^2 - \frac{\beta r_2}{K_2} N_1 N_2. \tag{2.4}$$

The fact that $f_1(N_1,N_2) = dN_1/dt$ and $f_2(N_1,N_2) = dN_2/dt$ make these ordinary differential equations. The equations provide the structure of the model, and specify the relationships between the parameters (r_1, r_2, K_1, K_2, α, and β) and the state variables (N_1 and N_2). The relationship between the parameters and each state variable can be expressed in other ways—for example, the per-capita change or the population size at time t, but only the functions $f_1(N_1,N_2)$ and $f_2(N_1,N_2)$ are ordinary differential equations. Thus the equations for the per-capita rates are not differential equations.

While both $f_1(N_1,N_2)$ and $f_2(N_1,N_2)$ are quadratic functions, the Lotka–Volterra equations are considered linear because the equilibrium solution is found by setting functions equal to zero and solving the resulting equations:

$$\begin{aligned} N_1 + \alpha N_2 &= K_1 \\ \beta N_1 + N_2 &= K_2 \end{aligned} \tag{2.5}$$

which are linear. Thus for theorists the term "linear" has a precise meaning about the form of the equations defining the solution, and the Lotka–Volterra models are a set of linear differential equations, regardless of the fact that the original functions are quadratic. Experimental ecologists are often less precise and discuss nonlinear relationships between parameters and variables, but this does not necessarily force the underlying model of differential equations to be nonlinear. For example, we could imagine that r_1 depends on densities in a nonlinear fashion [i.e. $r_1 = g_1(N_1,N_2)$], but the model is still linear because r_1 cancels out and is not a part of the equations defining the solution.

Once the equilibrium points have been determined, linear asymptotic stability analysis is then used to decide if the points are stable, neutral, or unstable and if the system cycles or not. Here the term linear refers to the way in which stability is analyzed because we assume near the equilibrium point that the differential equations, e.g. $f_1(N_1,N_2)$ and $f_2(N_1,N_2)$, are linear with respect to changes in N_1 and N_2 (see May 1973, Vandermeer 1973). Stability is determined by deciding if the populations return to the equilibrium point or not when they are perturbed just slightly off the equilibrium point.

There are several things to keep in mind. Parameters, for example the *r*s and *K*s in the competition model, define how various state variables (e.g. species densities) interact with each other. In systems with more than one equilibrium point, the different compositions of species found at the different equilibria are governed by the same set of parameters. These parameters do not change, and whether a system has multiple stable equilibria or not is simply determined by the parameter values. Moreover, the presence or absence of one or more species at a particular equilibrium point does not define a different model structure, just a different equilibrium point. From this viewpoint, alternative states occur in dynamical systems when multiple equilibria exist for the same set of parameter values.

If we add a parameter, then we create a different model. For example, adding a parameter for an Allee effect to a logistic growth model creates a new model. While it can be informative to examine the effects of adding parameters, we must remember that once we do so we are comparing two different models.

Experimental and theoretical ecologists often mean slightly different things when talking about perturbation to a system. For theorists, perturbations to a model means moving the state variables—species densities—off the equilibrium point and following the response without further intervention. Experimentalists call this kind of perturbation a "pulse" (Bender et al. 1984), and experiments in which species are set at initial conditions and followed over time are typical examples. The long-term plots of Tilman et al. (2001), in which different competing plant species were sown in different combinations followed for 7 years is good example of an experiment using a pulse perturbation. It is much less common for experimental ecologists to use pulse perturbations in a community that is at or near an equilibrium point—for example, reduce the density of a competitor in an established community and let the system run free. One attempt of this sort (Petraitis et al. 2009), which was done as a test for multiple stable states, will be discussed in more detail in Chapter 4.

Experimentalists also use press perturbations, in which the densities of one or more species are held constant. Typical examples of press perturbations include experiments in which predators are either excluded or included by cages. However, by holding densities constant, we are changing a state variable into a parameter that defines a constant effect. A cage that excludes a predator holds the density of the predator constant: this fact reduces the two equations for the predator–prey model to one equation for the prey in which the predator density becomes a parameter. As we shall see later, many of the models that have been used as examples for multiple stable states have taken advantage of this approach and have examined interactions between two species—for example vegetative biomass and a grazer—as a one-species model.

In much the same vein, experimental ecologists often distinguish between intrinsic and extrinsic perturbations and between perturbations of state variables and parameters. Experimentalists usually consider any event from outside the system that affects the species under study as being an extrinsic disturbance. Fires, storm damage, flood events, and pest outbreaks often fall into this category. Yet for models of ecological processes, intrinsic versus extrinsic and parameters versus state variables are defined by the model itself. If such events are included in a model as parameters (e.g. mortality due to fire) or as state variables (e.g. density of a pest), then these events are intrinsic to the model. However, if fire and pest outbreaks are not part of the model, then there is no parameter defining their effect on state variables. Thus the only way to envision the effects of those events that are outside the structure of the model is to view them as pulse perturbations of state variables. While ecologists might imagine shifts in parameter values, such shifts can only occur if the parameter was included in the model to begin with.

Part of the confusion about how perturbations can affect parameters and state variables arises from the use of the ball and cup analogy to understand equilibrium points and stability. While ecologists often cite Lewontin (1969) for this extraordinarily useful way to visualize multiple stable states, Lotka (1956) discussed this analogy much earlier. In the simplest case, state variables are represented as a ball and parameters "define" the shape of a surface. If the system has one stable equilibrium point, the surface is a cup and the ball rests at the bottom. If the ball is displaced, it always rolls back to the bottom of the cup, which represents the stable equilibrium point. If the system has one unstable equilibrium point, the surface is an inverted cup. The ball can sit on highest point, but the slightest disturbance will cause the ball to roll off that point. If there are multiple stable states then the surface will have hills and valleys, and the ball—the state variables—can be trapped in any one of valleys, which are often called basins of attraction. It is commonly thought that parameters alone define the shape of the surface (e.g. Beisner et al. 2003), but this is true only in the cases with one state variable. As we shall see in Chapter 8 as part of our discussion of common misconceptions, the shape of the basins of attraction also depends on the path traversed by the state variables when perturbed from an equilibrium point.

Finally, it must be remembered that in the classical models of species interactions used by ecologists, parameters are abstractions that simplify the links between ecological processes and rates of change. For example, in the logistic growth model r and K define the rate of change of a single species as a function of a stable variable, the density N. Yet the link between population parameters and an individual's probability of mortality and expected fecundity require assumptions about how the relationship between the parameters r and K and the state variable N summarizes the linkages of the "average" individual's time and energy budget to population-level processes. The per-capita form of the logistic model, for example, could easily be used to describe the relationship between the average speed of the typical car on the highway (a per-capita rate) as a function of the number of cars per 100 m of highway (a density measure).

However, the equation cannot be used to separate out cause and effect without making assumptions about the underlying relationships. Do cars slow down because there are more cars on the highway or are there more cars on the highway because drivers ahead are slowing down to look at an accident? For ecological models we make the same sort of implicit assumptions about mechanisms (Dunham and Beaupre 1998). Birth rates do change because of density per se, but because we assume resources, which are converted into offspring, are in short supply when density is high. We make certain assumptions about the mapping of offspring onto resources and resources onto density so that we can describe birth rate in terms of density.

2.2 Allee or dispensation effects and multiple stable states for a single species

A single-species population growth model with an Allee effect is the one of the simplest ecological examples of a system with multiple stable points. The typical representation of the Allee effect is a point of the rate of population growth (dN/dt) versus density (see Figure 2.1). The stable points are zero and K with an unstable point at N_0. If the population falls below N_0 it will go extinct and will not be able to recover unless re-introduction efforts push the founding population above N_0. The usual verbal explanation for the Allee effect based on social interactions among individuals is that minimum population size is needed before social interactions can counteract the effects of low population size on birth and death rates. The loss of the passenger pigeon is a commonly cited example of this sort of social facilitation. More generally, the wavy curve associated with an Allee effect is known as dispensation (Clark 1976). Loehle (1989a) notes that in forests, aggregation effects—for example better recruitment of seedlings near the adult canopy—can give rise to dispensation and result in multiple stable states.

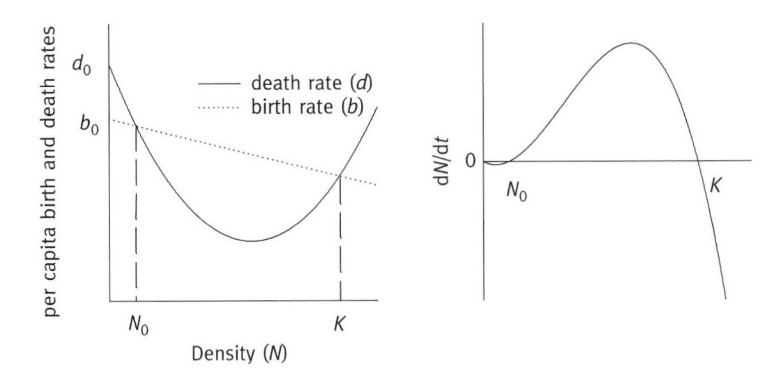

Figure 2.1 The Allee effect shown as the combined effect of birth and death rates and as rate of population growth. This is a classic example of an Allee effect where the death rate is high at low densities, drops as densities increase, and then climbs with further increases.

For the Allee effects, there are at least 11 different models that have been proposed as descriptors of this pattern (Boukal and Berec 2002). One of the simplest involves adding a third-order term to the logistic growth curve, which gives

$$f(N) = \frac{dN}{dt} = rN\left(1 - \frac{N}{K} - \frac{aN^2}{K}\right). \tag{2.6}$$

The equation reduces to the standard logistic growth model when $a = 0$, and shows an Allee effect when the effect of the third-order term is positive (i.e. $a < 0$). It is worth noting that Lotka (1956) suggested that the rate of population growth could be modeled as a polynomial function and that ignoring higher-order terms beyond the quadratic term provided the simplest model of population regulation.

While this model is a useful description of population growth in a population with an Allee effect, the ecological meanings of r, K, and a are not clear. We do not measure parameters such as r, K, and a directly, but rather birth and death rates. So it is worthwhile to step back and ask—what is the relationship between this type of model and birth and death rates?

Let's start with the logistic growth model. Logistic growth assumes that the per-capita rate of population growth (dN/Ndt) is linear with density, and if we rewrite the per-capita rate in terms of births and deaths,

$$\left(\frac{1}{N}\right)\left(\frac{dN}{dt}\right) = b(N) - d(N), \tag{2.7}$$

this means that the difference between births and deaths, i.e. $b(N) - d(N)$, is linear. The birth and death functions themselves need not be linear, only the difference. Even so, one common—and the easiest—way of representing the difference is to make the birth and death functions linear. Suppose

$$\begin{aligned} b(N) &= b_0 - uN \\ d(N) &= d_0 + vN. \end{aligned} \tag{2.8}$$

The minimum rates for births and deaths are b_0 and d_0 (i.e. the intercepts), and the linear density-dependent changes in these rates are the slopes v and u. By rearranging terms, we find $r = b_0 - d_0$ and $K = (b_0 - d_0)/(v + u)$. The maximum rate of increase r makes intuitive sense and is the maximum difference between births and deaths, which occurs at $N = 0$. In contrast, the carrying capacity K confounds the effects of rate of approach to K (i.e. effects of v and u) with r. This problem arises from our assumptions about how density affects births and deaths, and other plausible assumptions will give us a different expression for K (e.g. Vandermeer and Goldberg 2003). There are two points to keep in mind. First, birth and death rates are determined by resources levels, the presence of predators and competitors, and so on, and not by densities directly. Second, this means density is a proxy of the real causes of variation in birth and death rates. Thus there are also additional implicit assumptions about the links between resource levels and densities, and between resources and the rates of births and deaths.

Now an Allee effect could be modeled most transparently by making either $b(N)$ or $d(N)$ a quadratic function. For example, suppose death rate is

$$d(N) = d_0 + vN + sN^2. \quad (2.9)$$

Here the term s gives the nonlinear effect of density on the death rate. Now imagine that the death rate at very small population sizes is much greater than the birth rate because there is no social facilitation to prevent high mortality (i.e. $b_0 < d_0$). The exact ecological cause could be due to various factors, and the social facilitation could affect either adult mortality rates (e.g. benefits of herding) or juvenile mortality rates (e.g. benefits of group guarding at natal sites). Regardless of the cause, the effects of facilitation among individuals increase as population size grows, and mortality rates decline. However, the declines due to facilitation are eventually overshadowed by the "pure" density effects of population density, and mortality rates begin to rise.

The curves for death and birth rates now cross at two points, N_0 and K (Figure 2.1) and give the familiar population growth curve for Allee effect (sensu stricto, the strong Allee effect; see Stephens et al. 1999). There is an additional requirement for a strong Allee effect, and that is that the two functions must cross at least twice—once at the conventional point of stable equilibrium, which is normally defined as K, and at least once below K. The most plausible scenario is that the death rate is much greater than the birth rate at low densities (i.e. $b_0 < d_0$), but this does not need to be the case. For example, it is perfectly possible to have a "weak" Allee effect at low densities (Stephens et al. 1999) and have positive growth rates ($dN/dt > 0$) for all values of N below K. The curve for dN/dt will be asymmetrical but will have only one equilibrium point, which is stable at K. Nonlinearity does not guarantee multiple stable points, and as we shall see later, multiple stable points can occur even in linear systems.

We could also imagine that a population with social facilitation could also change from having the classic Allee effect of two nonzero equilibrium points (N_0 and K) to having a single stable point with shifts in environmental conditions. Suppose birth rates are dependent on resource availability and thus the birth curve moves upwards with improved condition. Imagine the shift is caused by changes in b_0, and so the birth curve is simply shifted up. Under very poor conditions the birth curve is always below the death curve, and there is no equilibrium point. Eventually the birth curve shifts upwards enough to touch the death curve at one point, which is unstable. Further increases results in three equilibrium points, the unstable point N_0 and the stable point at zero and K. The internal points N_0 and K diverge with improved conditions and finally when $b_0 > d_0$, the unstable point N_0 disappears and the stable point at zero becomes unstable.

Species that may have Allee effects have been offered as examples of multiple stable states in nature. One of the more widely cited examples is the shift from coral to algal dominance with the loss of the sea urchin *Diadema antillarum*. Under strong grazing pressure of algae by *D. antillarum*, the Caribbean was mostly dominated by corals.

Diadema antillarum suffered catastrophic mortality due to a disease outbreak in 1983–4 and there has been a prolonged lack of recovery (Lessios 1988). Many areas in the Caribbean have become covered by algae. Knowlton (1992, 2004) suggested that the lack of recovery was due to an Allee effect, and *D. antillarum* may have a two stable equilibria—one at high population density and the other at low population density.

2.3 Consumer models with one, two, and four equations

Many models of multiple stable states for two species are grazing models in which the second species—the grazer—is added as a parameter in a single-species model. May's (1977) presentation of Noy-Meir's (1975) model of grazing by range animals and their effects on vegetation is one of the earliest, best known, and most widely imitated examples. May (1977) assumed that the rate of growth of vegetation is $g(N)$ where N is vegetative biomass and the rate of removal by grazers is $ph(N)$ where p is the number of grazers and $h(N)$ is the per-capita rate of consumption. We use p rather than P to denote grazers because number of grazers is a parameter, not a state variable. The rate of change in vegetation is then

$$\frac{\mathrm{d}N}{\mathrm{d}t} = Ng(N) - pNh(N). \tag{2.10}$$

Now assume vegetation is self-limiting and modeled as logistic growth. Thus $g(N) = r(1 - N/K)$. Note that $Nh(N)$ is the rate of biomass consumption per grazer and thus is the functional response. Both Types II and III can be modeled as a Hill function (Scheffer et al. 2001), which is

$$Nh(N) = \frac{mN^q}{w^q + N^q}. \tag{2.11}$$

When $q = 1$, the function reduces to the Michaelis–Menton equation and is the familiar Type II functional response curve (Figure 2.2). As N becomes large, the rate of consumption saturates at m, and at half-saturation, $N = w$. When $q > 1$, the function takes on the characteristic S shape of a Type III functional response (Figure 2.2). Parameter q adjusts the steepness of the threshold and larger values will make the curve steeper. Consumption still saturates at m with w as the half-saturation point, which is at the inflection point of the curve.

The number of grazers is controlled by managers and so the total rate of removal of vegetative biomass depends on the number of grazers or stocking density. If only a few animals are allowed on the range, then the rate of removal of biomass is low. As more animals are allowed to graze, the rate increases. The system will be at equilibrium when $\mathrm{d}N/\mathrm{d}t = 0$, which occurs when $Ng(N) = pNh(N)$. Figure 2.2 illustrates how if consumption follows a Type II functional response, the system can move from one stable equilibrium point to two equilibrium points as stocking density is increased. The lower point is unstable, and if $\mathrm{d}N/\mathrm{d}t$ was plotted against N, the plot would be

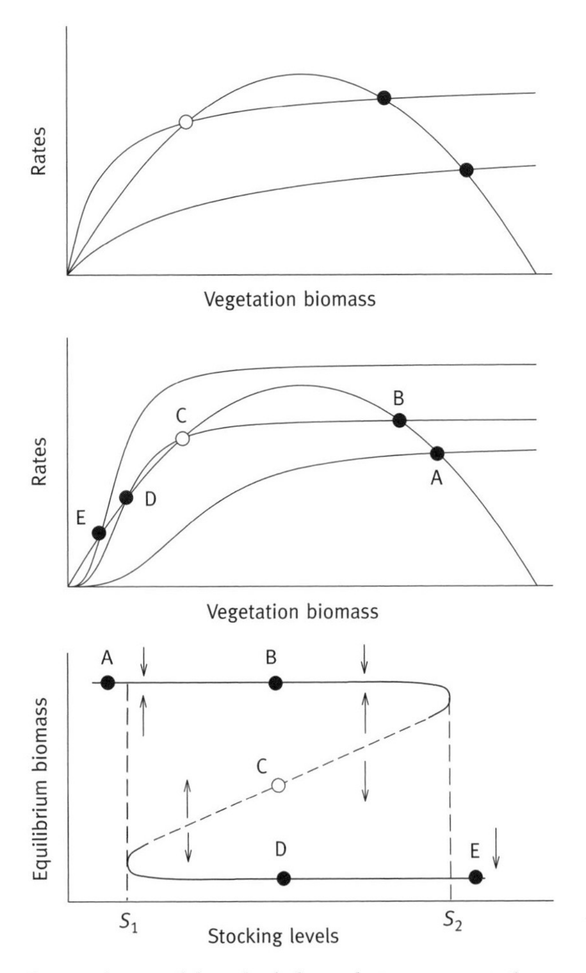

Figure 2.2 Single-species grazing model as the balance between vegetative growth and harvesting. Hump-shaped curves in the top two panels are the rate of accumulation of vegetative biomass. The top panel gives harvesting curves with a Type II functional response; the middle panel shows the effects of a Type III functional response. The bottom panel shows a plot of equilibrium points. Solid circles are stable nodes; open are unstable. Redrawn from: May R. M. (1977). Thresholds and break-points in ecosystems with a multiplicity of stable states. *Nature* 269, 471–477. Reproduced with permission from Nature Publishing Group.

indistinguishable from an Allee effect shown in Figure 2.1. This underscores May's (1977) suggestion that theory is often at best a metaphor for ecological processes, and our models will be only as good as the assumptions we use to build them.

Using a Type III functional response opens the possibility of there being multiple stable states (Figure 2.2). At low stocking levels, there is one stable point (point A in Figure 2.2) but as stocking levels increase the system abruptly develops three equilibrium points—two stable and one unstable (e.g. points B, C, and D). The unstable point is initially close to the lower stable point, and it would be very difficult to detect a difference in vegetative biomass at the two equilibrium points. As stocking levels are increased further, the unstable point diverges from the lower stable point and moves towards the upper stable point. Finally, the unstable point and the upper stable point merge and disappear so there is an only single stable point (e.g. point E).

The changes in the number and type of equilibrium points with different stocking levels can be represented as either as a ball on a surface with hills and valleys or as an S-shaped curve. Figure 2.3 shows three cross-sections of the hills and valleys. At the extremes of low and high stocking levels there is only one valley, but at intermediate levels there are two valleys. If the hills and valley are projected down as a plot of the equilibrium values for vegetative biomass against stocking levels, we then get the familiar S-shaped curve (i.e. as in Figure 2.2). The solid lines give the stable equilibrium points and the dotted line is a ridge line, which separates the two basins of attraction. This ridge line is May's (1977) "breakpoint curve." There are also two discontinuous jumps at the turns or folds in the S curve and a sudden shift between the stable equilibrium points. As the stocking changes and a fold is approached, the distance

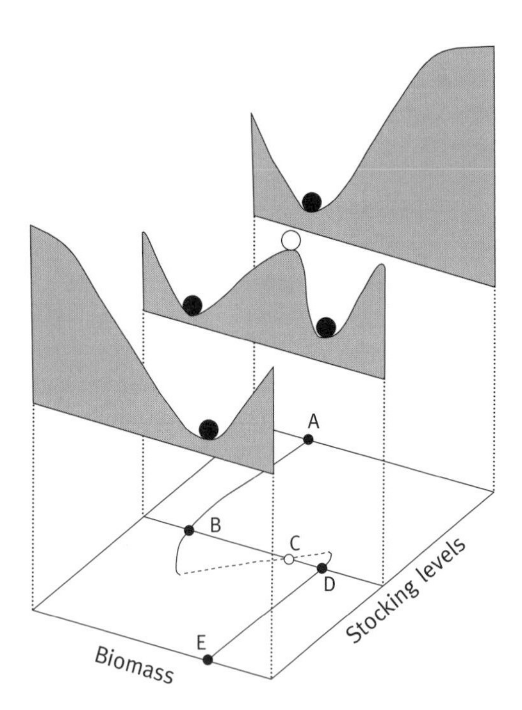

Figure 2.3 The changes in the number and type of equilibrium points with stocking density as a ball and cup diagram. The S-shaped curve is shown as a projection of the peaks and valleys. The letters and shading of the balls match usage in Figure 2.2. Redrawn from Scheffer, M., Hosper, S. H., Meijer, M. L., Moss, B., and Jeppesen, E. (1993). Alternative equilibria in shallow lakes. *Trends in Ecology and Evolution* 8, 275–279. Reproduced with permission from Elsevier.

between one of the stable points and the ridge line shrinks. Thus, close to the folds, small perturbations in vegetative biomass that are unrelated to stocking levels may tip the system over the breakpoint curve and cause the vegetative biomass to shift abruptly to the other stable point. Many ecologists used the term "threshold" for both crossing the breakpoint curve and the discontinuous jumps at the turns. This has led to quite a bit of confusion, which we will attempt to clear up in our discussion of misconceptions about multiple stable states in Chapter 8.

The plot reveals several hallmarks of systems with multiple stable states as envisioned by ecologists (May 1977). First, the range of parameters below the two folds of the S-shaped curve is known as a bifurcation set because a single equilibrium point abruptly splits or bifurcates into three—two stable and one unstable—at the folds. The existence of a sudden jump at the folds is often taken as strong evidence for the existence of multiple stable states in nature. Smooth and continuous changes in a parameter produce discontinuous shifts in the equilibrium values of the state variable. While May (1977) is usually credited with bringing this phenomenon to the attention of ecologists, the existence and mathematics of discontinuous jumps were well understood in the 1960s (Arnol'd 1992).

Second, folds occur at two points—one at each stocking levels, S_1 and S_2—and at these points the system shows catastrophic shifts in biomass with very small changes in stocking density. The system becomes more sensitive to small changes in stocking levels as the folds are approached. Thus very sudden shifts in species composition with small changes in environmental conditions are also considered an indicator of multiple stable states in nature, but as we shall see later this needs not be the case. Finally, once these folds are passed and the system shifts, it cannot easily return to the "original" state.

Hysteresis can be detected by perturbing state variables (e.g. population densities) or changing parameter values (e.g. per-capita birth rates). Hysteresis does not mean that future events are random, stochastic, or unpredictable as has been often incorrectly inferred from May's (1977) reference to history and Cleopatra's nose. Hysteresis means that we must know both the state variables and the parameter values to predict the future course of the system.

It is important to note that the models considered so far involve only one differential equation. The Noy–Meir grazing model is a single equation for vegetation, and while grazers are a second species in this system, the effect of grazer density is modeled as a parameter. Many of the well-known two-species models for multiple stable states are, in fact, of this form with the second species introduced as a parameter, not a state variable (e.g. May 1977, Scheffer et al. 1993, 2001, 2003, van de Koppel et al. 1997, 2001).

We can extend the grazing model to the more familiar predator–prey model by adding a second equation for the predator, and the pair of equations is:

$$\frac{dN}{dt} = Ng(N) - PNh(N)$$
$$\frac{dP}{dt} = -dP + ePHh(N) \tag{2.12}$$

where d is the predator's death rate and e is the conversion of prey into predators. As in the grazing model, $g(N)$ is the growth rate of the prey and $h(N)$ is the harvest rate, and under the classical Lotka–Volterra model these functions are constants. The density of the predator is now P rather than p since it is a state variable and not a parameter. Rosenzweig and MacArthur (1963) showed a humped prey isocline is the result of logistic growth for the prey and a Type II functional response (Figure 2.4). The predator isocline is always a vertical line and the system has one equilibrium point.

Both the predator and the prey isoclines must become curvilinear in order to change the system to one with multiple stable states with both predator and prey present (Figure 2.4). First, the prey isocline must have a bend, which would develop if the growth rate of the prey includes an Allee effect. Second, the predator isocline must also bend sharply so that it crosses the prey isocline at three points. The predator isocline can be made to bend to the right by making the predator death rate or the harvest rate dependent on predator density. We could imagine that interactions among individual predators could either increase death rates or make capturing prey more difficult (i.e. mutual interference; Hassell 1978). However, a sharp bend requires the density dependence of either the death rate or harvest rate have a steep step-like shift—for example, a Hill function with a very large value for q. It seems more plausible that

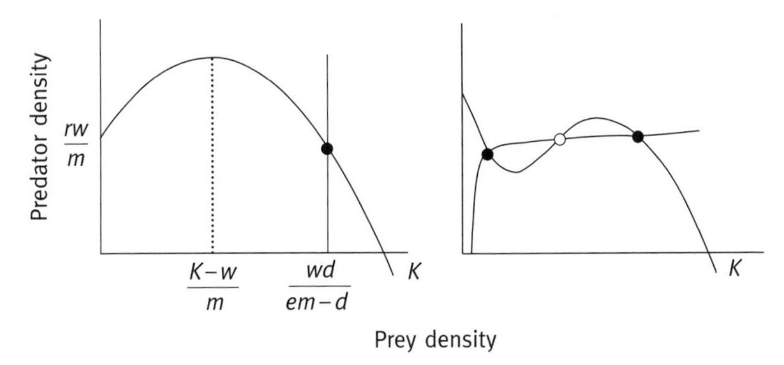

Figure 2.4 Isoclines for two predator–prey models. The left panel shows isoclines (solid lines) for logistic growth in prey and the Type II functional response. Intercepts and the position of the hump (dotted line) are defined in terms of the model parameters; w and m are parameters in the Hill function (see Equation 2.11), d is the per-capita death rate for the predator, e is the conversion rate of prey into predator (see Equation 2.12), and r and K are are parameters of the logistic growth model. The right panel shows the prey isocline with the Allee effect and the predator isocline with mutual interference among predators. This system has three stable points: two stable (solid dots) and one unstable (open dot).

strong inference among individual predators would be the underlying cause of such a sharp bend in the predator isocline.

The two stable equilibrium points differ in their stability properties. Perturbation of either the prey or predator from the lower point will cause the prey and predator to cycle back to the equilibrium point, but at the upper point, neither species shows oscillatory behavior. This is a good example of one of the ways in which the cup and ball analogy fails to capture the full dynamics of multiple stable states.

Field observations of predator-prey interactions might seem quite unusual and counter-intuitive if the underlying dynamics of the system were not fully understood. In this model, the equilibrium density of the predator is virtually the same at both stable points, yet prey levels and dynamics are quite different. At the lower stable point, it would appear as if the predators were very effective keeping prey density low, but perturbations could easily cause oscillations that would carry the system to the upper equilibrium point or drive the predator to extinction. In contrast at the upper point, the prey are much more common suggesting the predator is not effective in controlling the prey yet perturbations of the system would not cause cycling of predator and prey numbers. Without some knowledge of the underlying causes, it would be difficult untangle the interactions between prey and predator.

Predator-prey models with multiple stable states have been extended to include multiple prey species, and Holt et al's (1994) model is one of the best known examples. Multiple stable states in this model depend on priority effects. Chase (1999) uses a variation of the Holt et al. model as the theoretical underpinnings of his experimental test of alternative stable states, which we will discuss in Chapter 4. Both Chase's and Holt et al.'s models are structurally the same as to an earlier model developed by Armstrong (1979). All three models give qualitatively the same results.

The model has one predator (*P*), two prey (N_1 and N_2) and one resource (*R*), and the dynamics of the system is described by four equations:

$$\begin{aligned} \frac{dP}{dt} &= Pf(N_1, N_2) \\ \frac{dN_1}{dt} &= N_1[g_1(R) - Ph_1(N_1, N_2)] \\ \frac{dN_2}{dt} &= N_2[g_2(R) - Ph_2(N_1, N_2)] \\ \frac{dR}{dt} &= s(R) - N_1c_1(R) - N_2c_2(R) \end{aligned} \tag{2.13}$$

where $f(N_1,N_2)$ is the growth rate of the predator as a function of prey densities, $g_1(R)$ and $g_2(R)$ are the growth rate of prey species 1 and 2 as a function of resource levels, $h_1(N_1,N_2)$ and $h_2(N_1,N_2)$ are the predator's harvest rate of species 1 and 2, $s(R)$ is the renewal or supply rate of the resource, and $c_1(R)$ and $c_2(R)$ are the consumption rates of the resource by species 1 and 2. Holt et al. (1994) and Chase (1999) used this basic model; Armstrong's version includes multiple prey species.

In a system with one predator, one prey species, and one resource, the phase diagram is three-dimensional, but Armstrong (1979) and Holt et al. (1994) realized that at equilibrium, the resource is tied up either as biomass of predator or prey or is free and available. Armstrong and Holt et al. used this insight to constrain the phase diagram to two dimensions, although in different ways. We will use Holt et al.'s convention in which they envision a mass-balance constraint (i.e. the MBC line). The MBC line defines the amount of resource in the system at equilibrium as a combination of available resource and predator density, which is a proxy for predator biomass (Figure 2.5). Since the total amount of resource is set, the amount of resource tied up as prey biomass is defined by the MBC line and does not need to be explicitly stated. At equilibrium, the system could have a large number of predators and low available resource or a few predators and lots of resource. If the total amount of resource in the system increases—that is, if the system becomes more productive—then the MBC line moves outwards (Figure 2.5A).

The equilibrium point can lie anywhere on the MBC line and so we need additional information about the zero net growth rate of the prey (ZNGI) to tell us where that

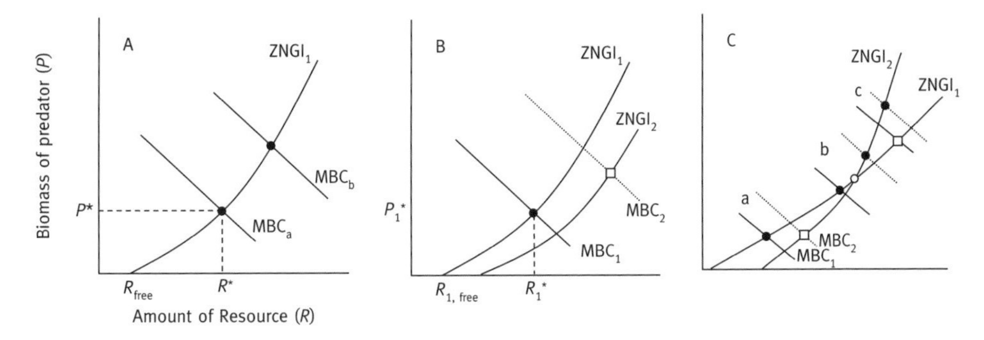

Figure 2.5 An adaptation of Holt et al.'s (1994) model of the effects of productivity on a system with two prey species and one predator. Axes are amount of resource (R) and the biomass of predator (P); letters with asterisks give the equilibrium levels. R_{free} is the amount of resource that is free in the system at equilibrium. Solid circles are stable points not open to invasion; open circles are unstable equilibrium points; open squares are stable points open to invasion by the other species. (A) Increasing productivity pushes the MBC line out (from MBC_a to MBC_b) and shifts the equilibrium point. (B) Two-species competition in which species 1 wins because its ZNGI intercepts the R-axis closer to the origin; species 1 drives the amount of free resource below levels that are needed to sustain species 2. Note that species 2 can be in equilibrium with the predator (the open square) but this is open to invasion by species 1 and displacement to the stable point (the closed circle). (C) Changes in productivity push the pair of MBC lines outwards and move the system through three distinct phases. In all three, the predator is present. At low productivity only species 1 occurs (pair a of MBC lines) while at high productivity only species 2 occurs (pair c). At intermediate levels there is an unstable point and priority effects (pair b).

might be. The system is at equilibrium where the MBC line and the ZNGI cross. The intercept of the ZNGI with the resource axis gives the amount of free resource at equilibrium if there are no predators in the system. The closer the ZNGI intercept is to the origin the more efficient the prey species is at acquiring and utilizing resources. An efficient consumer will drive down the equilibrium level of available resource. In a system of two consumers and no predator, the better resource competitor wins, i.e. the species whose intercept is closer to the origin (Figure 2.5B). The slope of the ZNGI gives the prey's susceptibility to predation. If there are few predators in the system at equilibrium then it takes less resource for a prey species to maintain zero net growth. If the number of predators increases, then more prey will be consumed and thus it will require more resource to maintain zero net growth. If a small increase in predator density has a big impact on the prey, then this should cause a large increase in available resource and the ZNGI will be very shallow.

Now consider a system with two prey species. Each prey species has its own ZNGI and MBC line. There are a number of different ways the ZNGIs and MBC lines can be drawn (Holt et al. 1994), but we will consider only the one that gives rise to priority effects and alternative stable states. This is the scenario for unstable coexistence of prey species in the presence of a predator (Figure 2.5C). For both stable and unstable coexistence, the ZNGI must cross, and for our example let us assume the ZNGI for species 2 is steeper than the ZNGI for species 1, and its intercept is farther from the origin. This means species 2 is a poorer competitor for resource but is less sensitive to the effects of predation. Now assume the MBC line for species 2 is always outside the MBC line for species 1. This implies that the biomass or density of species 2 is less than that of species 1 at equilibrium. Under this scenario, the system can only have one prey species in equilibrium with the predator at an particular level of resource.

The pair of MBC lines will move together and outwards as productivity levels increase, and this determines which prey species occurs with the predator. At low levels of resource, both MBC lines intersect the ZNGIs below where the ZNGIs cross (two lines in set A) and at equilibrium the system will contain the predator and only species 1. At high levels of productivity (set C), species 2 wins and coexists with the predator. At intermediate levels, the pair of MBC lines straddles the point where the ZNGIs cross and this is the region in which priority effects matter (set B). Which prey species enters the system first and gains a foothold determines the outcome. In some cases only species 1 coexists with the predator and in others only species 2 coexists with the predator.

2.4 Competition models

The Lotka–Volterra competition model for two species with a saddle node is the simplest two-species model in ecology with multiple stable states (Knowlton 1992, Lotka 1956). The saddle node is an unstable point between two stable points. At each stable equilibrium point, one species excludes the other. Species 1 at equilibrium cannot be invaded by species 2, and vice versa. The saddle node is unstable but both species are "at equilibrium." Any perturbation of the system that shifts densities away from the saddle node will cause the system to move to one of the stable points. Replacement of one species by the other can only occur if a disturbance that either reduces the density of the current dominant or allows invasion by the other species so that the species composition shifts past the separatrix, which is the line that defines the boundary between the two basins of attraction (Slobodkin 1961). Not only do initial conditions of the state variables (i.e. the initial densities of the two species) determine which species wins but they also give rise to priority effects in which the order of and timing between arrival of species determines the final outcome. Park's (1948, 1954) work with *Tribolium*, while not usually considered an example of multiple stable states or priority effects, is one of earliest experimental demonstrations of how initial conditions determine the outcome of competition between two species. Finally, and most importantly, this simple model clearly illustrates Lewontin's (1969) comment that while multiple stable points are possible for linear models, one or more species must be missing at each stable point. The existence of a saddle node requires that the inequalities $\frac{1}{\alpha}<\frac{K_2}{K_1}<\beta$ must hold, and this requires the interactions between species to be very asymmetrical. In addition, the inequalities imply that either a or b must be greater than 1, and for one of the species interspecific effects must be greater than intraspecific effects. This can be seen by rewriting Equation (2.3)

$$\frac{dN_1}{N_1 dt} = r_1\left(1 - \frac{N_1 + \alpha N_2}{K_1}\right) \tag{2.14}$$

The term $(N_1 + \alpha N_1)/K_1$ gives the combined limiting effect of species 1 and 2 where the competition coefficient α is the correction factor for converting individuals of species 2 into an equivalent number of species 1. The equation reduces to the logistic growth model if $\alpha = 0$. When α is greater than one, the impact of an individual from species 2 is greater than the intraspecific effect. It is commonly thought that this should be rare in nature since intraspecific competition—particularly for resources—is usually much greater than interspecific competition. However, α can easily be greater than one under interference competition or in situations where a competitor is also a predator.

It is also not commonly appreciated that linear models can show hysteresis-like behavior, and it turns out that we can make a plot of equilibrium densities against environmental conditions for the classical Lotka–Volterra model that looks like it has two folds. Suppose changes in environmental conditions can increase the carrying capacity of species 1 (K_1) but has no effect on species 2. This could be simply an additional resource that can be utilized by species 1 but is unavailable or cannot be used by species 2. In a relatively poor environment for species 1, the isocline for species 1 is completely inside the isocline for species 2, and species 1 can resist any colonization attempt by species 2 (Figure 2.6). As conditions improve for species 1, its equilibrium density, which is K_1, increases, and the isocline for species 1 shifts outwards. Eventually the isocline for species 1 touches and crosses the isocline for species 2 at the point

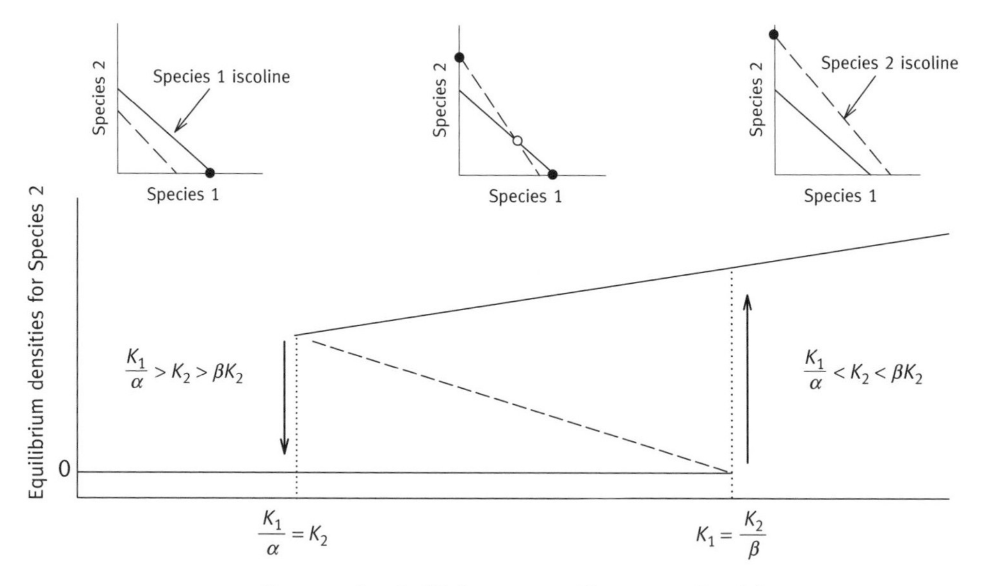

Figure 2.6 Standard Lotka–Volterra model for two-species competition showing a Z-shaped fold with changes in resource levels (value of K_2). The three phase diagrams above show species isoclines at three points along the resource gradient. Below, the Z-shaped curve shows the equilibrium points for species 2 (stable as solid line and unstable as dashed line). Redrawn from Petraitis, P. S. and Hoffman., C. (2010). Multiple stable states and relationship between thresholds in processes and states. *Marine Ecology–Progress Series* 413, 189–200. Reproduced with permission from Inter-Research Science Center.

(0, K_1). Once this occurs, the system has three equilibrium points—two stable and one unstable. Further improvements in conditions will then push the isocline for species 1 outside the isocline for species 2 and at that point, species 1 can resist all attempts by species 2 to invade the system.

Figure 2.6 shows the equilibrium densities and breakpoint ridge for species 2 plotted against environmental conditions. The curve is Z-shaped and is reminiscent of the S-shaped curve in Figure 2.2. The solid lines give the stable equilibrium points (i.e. 0 and K_1) and the dotted line shows the unstable node or the breakpoint ridge. However, this is not hysteresis in the strict sense, even though the ability of species 1 to invade and resist invasion by species 2 shows lags behind changes in environmental conditions. The system is linear, not nonlinear, and so does not fall within the class of mathematical models that define hysteresis. Second, and more importantly, the ability of species to invade the switch between the two stable equilibrium points—one for species 1 and the other for species 2—depends on an outside pool of colonists that are not part of the model. Invasions are perturbations to, not parameters of, the model.

The classical Lotka–Volterra model can easily be altered so that there are three equilibrium points with both species present. All that needs to be done is to change the isoclines in the phase diagram from straight lines to curves that cross three times. The isoclines will curve whenever a nonlinear term is added (e.g. Gilpin and Ayala 1973). Vandermeer (1973) gives a variety of examples with a verbal explanation of why the isoclines might curve and cross more than once. Walter et al. (1981) provides an explicit example of a phase diagram of grasses versus woody plants with curved isoclines in their review of stability in grassland–woodland ecosystems.

A common approach to create competition models with multiple stable states is to introduce a term with a steep step-like form (Petraitis and Dudgeon 2004). They assume that two species are competitors for space in an open system:

$$\begin{aligned}\frac{dN_1}{dt} &= c_1 - d_1N_1 - z_1N_2 \\ \frac{dN_2}{dt} &= c_2 - d_2N_2 - z_2N_1\end{aligned} \tag{2.15}$$

where c_1 and c_2 are recruitment rates, and mortality from recruitment to adulthood is due to either conspecifics (i.e. d_1 and d_2) or interspecific competition (z_1 and z_2). Now suppose that at least one process shows a steep threshold over a range of parameter values. For example, suppose the adult replacement rate for species 1 includes a term for enhancement of recruitment due to the presence of adults, and so

$$\frac{dN_1}{dt} = c_1 + g(N_1) - d_1N_1 - z_1N_2 \tag{2.16}$$

and let $g(N_1)$ follow a Hill function (Scheffer et al. 2001):

$$g(N_1) = \frac{mN_1^q}{N_1^q + w^q}. \tag{2.17}$$

If q is large enough, recruitment will have a steep critical threshold with adult density, and the system will have an S-shaped curve. Adults may increase recruitment via social facilitation by preventing juvenile mortality. Petraitis and Dudgeon (2004) noted that the presence of adult macroalgae often prevents zygotes from being washed out of adult stands and helps limit juvenile mortality due to desiccation. This benefit, however, may be an all or nothing effect, and there may need to be a critical minimum number of adults in order for there to be an effect.

This critical number of adults may also depend on environmental conditions. For macroalgae, only a few adult plants may be needed to prevent loss of recruits in areas with little wave surge, and thus the position of the threshold, which is determined by w, would be to the left. In contrast, the required number of adults in exposed areas may be much larger where more adults might be needed to confer the same amount of improvement in recruitment. Thus w would be larger and the threshold would be shifted to the right.

If the threshold is steep enough, there will be two stable states over a range of intermediate values for w (Figure 2.7). Species 1 is common and species 2 is rare when the threshold for a jump in recruitment occurs in the presence of a few adults of species 1 (i.e. small w). As environmental conditions change so that more adult plants are required to cross the threshold in recruitment (w increases), the system develops three internal equilibrium points—two stable and one unstable. Further increases in w eventually cause the system to return to having only one stable equilibrium point, but at which species 1 is rare and species 2 is common.

The presence of a steep threshold for a model parameter does not guarantee that a system will have multiple stable states, and it is possible to have moderately steep

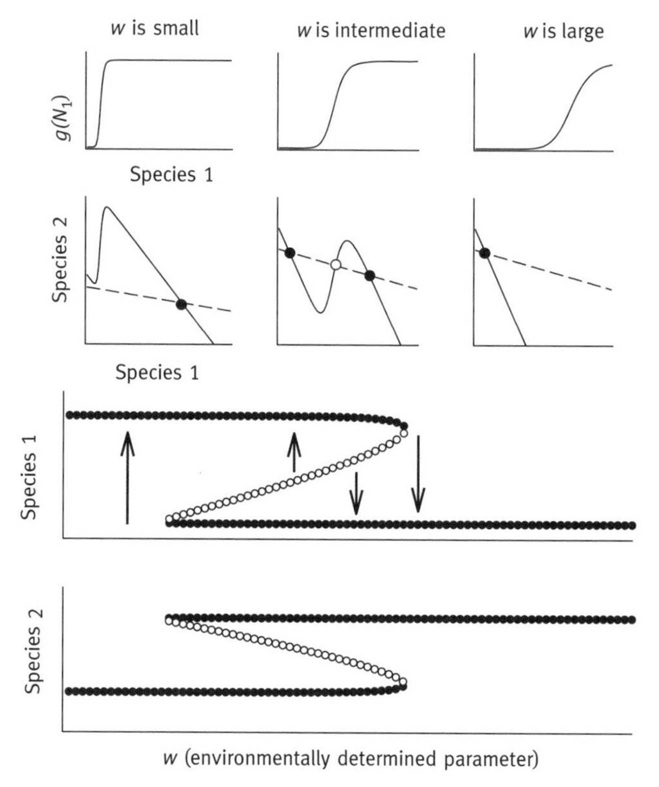

Figure 2.7 How changes in a parameter that is controlled by environmental conditions (w) can create an S-shaped curve in a model of competition between two species. Recruitment of species 1 shows a steep threshold that shifts with environmental conditions. The steep threshold causes the isocline for species 1 to be curved, and as the threshold shifts, the number of equilibrium points changes. Figure redrawn from Petraitis, P. S. and Dudgeon, S. R. (2004). Detection of alternative stable states in marine communities. *Journal of Experimental Marine Biology and Ecology* 300, 343–371. Reproduced with permission from Elsevier.

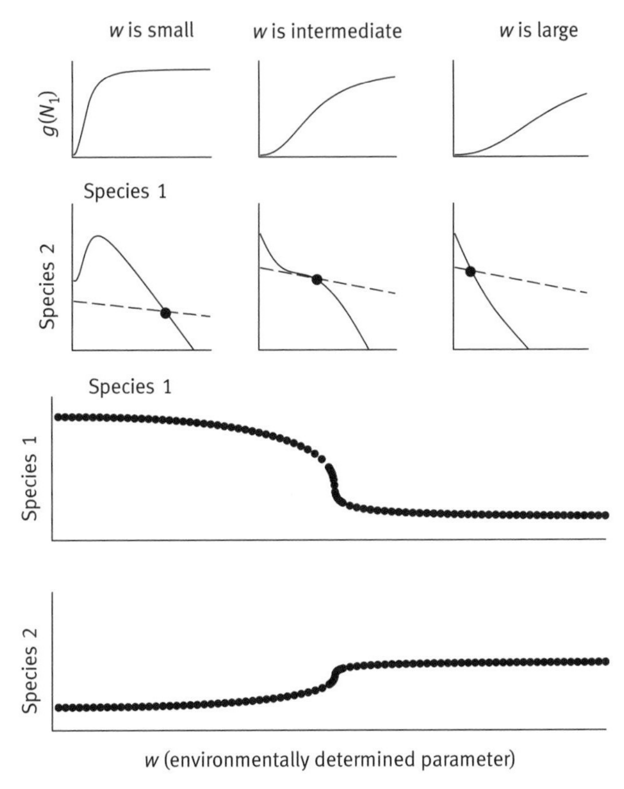

Figure 2.8 Thresholds in parameters do not ensure multiple stable states. The model here is identical to the model in Figure 2.7 but with a less steep threshold in recruitment of species 1. As a result, equilibrium densities show a smooth phase shift and do not have multiple stable states. Figure redrawn from Petraitis, P. S. and Dudgeon, S. R. (2004). Detection of alternative stable states in marine communities. *Journal of Experimental Marine Biology and Ecology* 300, 343–371. Reproduced with permission from Elsevier.

thresholds but only a single basin of attraction (Figure 2.8). We will return this point in Chapter 8 and provide a counter example of a model with multiple stable states but without thresholds in parameters. Finally, we must also remember that basins of attraction will change in shape and location with changes in parameters, but different equilibrium states under different sets of parameters cannot be considered alternative stable states. Thus demonstrations of abrupt shifts in species composition with small changes in environmental conditions (i.e. parameters) are not sufficient tests for multiple stable states.

2.5 Conclusions

It is quite easy to create plausible models that have multiple stable states. While models with multiple stable states are often nonlinear, but this need not be the case, as we have seen with the classical Lotka–Volterra model for competition between two species. Linear models that exhibit multiple stable states, however, will not have all species present at all equilibrium points (Lewontin 1969). There will always be one or more species missing at each equilibrium point. In nonlinear models, all species can be present at all equilibrium points, albeit with different densities at each point. For models to be nonlinear, one or more underlying process—such as births, deaths, and grazing pressure—must show a nonlinear change with density. Such nonlinearities make it possible, but not certain, that there will be multiple stable states. We will show in Chapter 8 that these nonlinearities need not be abrupt thresholds in order to have multiple stable states.

The S-shaped curve so commonly associated with the notion of multiple stable states gives rise to hysteresis. Systems with hysteresis have two characteristics. First there will be a sudden discontinuous jump from one stable point to another as a parameter changes and passes a critical value. Second, and more importantly, the system will not recover to the original stable point with the reversal in the parameter and backtracking across the critical value. Both linear and nonlinear models with multiple stable states will show these sudden shifts in equilibrium points and resistance to reversal, and so the presence of hysteresis is often thought to provide a strong test for multiple stable states. For linear models, these patterns are not hysteresis in the normal sense of the term. One or more species must be missing at each equilibrium point in linear models. Thus there must be an input of migrants to recolonize the system and to initiate the shift to an alternative stable state. Since the input of migrants is an extrinsic perturbation to the system, this is not hysteresis in the strict sense. Even so ecologists often refer to the resistance to recovery in dynamical models with extrinsic recruitment as hysteresis (e.g. Paine and Trimble 2004).

As we will see in Chapter 6, the presence of hysteresis depends on the amount of change in parameters—that is environmental conditions—relative to the amount of extrinsic noise in state variables such as densities. If densities are pushed around by random perturbations that are unrelated to changes in parameters then it is possible for the system not to show hysteresis.

Our brief examination of models of one, two, and three species suggests tests for multiple basins of attraction will provide the strongest tests for the existence of multiple stable states. However, the complexity of the basins of attraction increases with the number of multiple stable states (Feudel 2008). In systems with two alternative states, the boundaries between two basins are smooth. However, as the number of stable states increase, the majority of the basins will have fractal boundaries. These boundaries can be very complex in shape and intertwined, and it is possible that all basins will appear to cover the entire space of the system. As parameters change, boundaries can change from smooth to fractal, fractal to smooth, or from one fractal to another.

Tests for hysteresis are weaker in that the presence of hysteresis is consistent with multiple stable states, but the lack of hysteresis does not preclude them. While others have suggested that finding thresholds in parameters with changing environmental conditions is also evidence for multiple stable states, we will show that thresholds in parameters are neither required nor ensure the existence of multiple stable states.

Keep in mind that stability analysis of dynamical models tells us nothing about what stability might mean in natural communities or how parameters of models could be linked to environmental conditions or characteristics of species. For example, it has repeatedly been suggested that the nonlinearities that might promote multiple stable states are associated with ecosystem engineers, environmental switches and stressful abiotic conditions (e.g. Didham et al. 2005, Knowlton 1992, Petraitis and Latham 1999, Wilson and Agnew 1992). However, dynamical models themselves offer little guidance about practical approaches.

3

Detection of multiple stable states

The discussion of the various models in Chapter 2 provides a conceptual context for the two broad categories of approaches that have been used to detect multiple stable states in both natural and laboratory settings. These approaches are tests for multiple basins of attraction and demonstrations of hysteresis. In the following sections we will explore the criteria needed for these two experimental approaches to be used as tests of multiple stable states. Construction of models that are plausible and consistent with observed patterns is very widely used as a test of multiple stable states. However, modeling efforts are rarely linked with well-replicated experiments that include proper controls, and so we will defer our discussion of the links between modeling and patterns until Chapter 9.

3.1 Tests for multiple basins of attraction: Peterson's criteria

Theory tells us that multiple stable states mean there are two or more stable points for a single set of parameter values. The stable states are separated by a ridge in phase space that defines the edge between two basins of attraction. If the model is correct and the parameters are constant over time, then any two stable equilibria and the ridge that separates them can be represented as horizontal lines over time (Figure 3.1).

The most plausible scenario for a shift from one basin of attraction to another would involve the removal of a dominant species or most of a community by a natural disturbance that then provides the opportunity for members of an alterative community to invade and establish (Petraitis and Latham 1999). While the ridge line can be crossed at any point, it is hard to imagine how members of an alternative community entering as juveniles or propagules could overwhelm and displace populations of large established individuals without the removal of at least some of them. In terms of Figure 3.1, we could imagine that we are tracking species 1 that is common in one community i (i.e. where N^*_{1i} is the equilibrium density of species 1 in community i) and rare in the presumed alternative community j (N^*_{1j}). To tip the system from state i to j, the density of species 1 must be reduced below the density defined by the ridge line between states i and j, which is perturbation b in Figure 3.1. If there are multiple stable states, the density of species 1 will diverge to N^*_{1b}. Note that perturbing the density of species 1 to be below its density at the unstable node (e.g. perturbation a) will not cause the system to diverge.

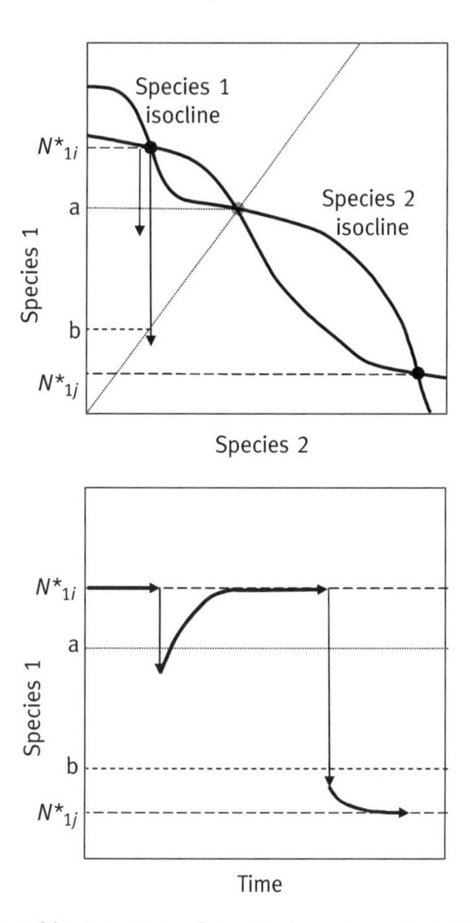

Figure 3.1 Shifts from one stable state to another state in response to perturbations. The top panel shows the phase diagram for two-species competition and the bottom panel shows changes in species 1 over time to two perturbations. The dotted line is where the first perturbation passes the density of species 1 at the unstable node and the short dashed line shows where the second perturbation passes the ridge line between the two basins of attraction.

Experimental evidence for multiple basins of attraction is considered to be the strongest test for multiple stable states, and many recent works have used an approach based on Peterson's criteria (Peterson 1984, Petraitis and Latham 1999). However, Peterson's criteria have often been misapplied, and so it is useful to quote in full Peterson's original statement. Peterson stated that "only by experiment could one convincingly demonstrate multiple stable states by showing that the very same site could come to be occupied by different, self-replicating communities." In addition, the experiment must use only pulse perturbations that mimic a natural event in spatial extent, temporal duration, and its effect on species in the system (Bender et al. 1984, Connell and Sousa 1983).

There are four separate requirements. Peterson clearly lays out that the experiment must be done in a single environment (i.e. at the same site), that the site is shown to have the potential to be occupied by two or more distinct communities, and that the communities are self-replicating, which was Peterson's proxy for stability. The fourth requirement was discussed in the exchange between Peterson and Connell and Sousa (1983), and is that the experimental manipulations must be pulse perturbations (sensu Bender et al. 1984) that mimic a natural event in spatial extent, temporal duration, and its effect on species in the system.

Peterson argued very forcefully and convincingly that the requirement of using "the very same site" is critical because using different sites can lead to incorrect assumptions about the environment in which the different community types occur. Multiple stable states require that two or more stable equilibrium points occur for the same set of parameter values, and ecologists usually assume the notion that the same set of parameter values implies the same environment. Lewontin (1969), for example, stated that multiple equilibria must occur in "a given habitat," and while he was quite clear that this meant the same set of parameter values for all equilibria, he acknowledged that in natural systems there may be noise in parameter values due to spatial and temporal variation.

The problem is that different sites may or may not be in the exactly the same environment with just a little bit of random noise. Often we think that several distinctly different species assemblages are multiple stable states because they occur as a mosaic of patches across what appears to be a uniform environment. However, it is possible that a patchwork pattern of communities could reflect an underlying pattern of undetected environmental variation, and thus each community occupies a unique environment. It is impossible to prove that different places in the seemingly uniform environment are in fact identical in every way because there is always a chance that important sources of environmental variation have been overlooked. Conversely, by using a collection of sites one might conclude that different communities in different environments are not multiple stable states when in fact they are. For example, dominant species within communities are often ecosystem engineers which can modify the environment (see Hastings et al. (2007) for a good review of ecosystem engineers). Thus different community states may appear to exist in different environments even though initial environmental conditions were identical.

Even though the requirement of using the "very same site" is compelling, most researchers using Peterson's criterion sidestep the requirement by using replicate sites in the same type of habitat. This approach is certainly more tractable experimentally, but unless more than one community state can be shown to occur at some of the sites, there is still the possibility that differences in community states at different sites are due to undetected environmental variation. This can a problem if habitats (i.e. the same environment) are defined in terms of species that occupy them (e.g. redwood forests and coral reefs), rather than by the range of environmental conditions found at a particular place and time (e.g. rocky intertidal shores and ephemeral desert pools).

Related problems arise in attempting to define different communities and pulse events, which are Peterson's second and third requirements. Operational definitions of different communities and natural perturbations require a priori knowledge about the amount of variation in species composition and disturbances within and between the presumed communities. It is important to realize that whether a perturbation is a pulse or press is defined by the system and its natural history. A cage that excludes consumers and is maintained for a long period of time but then removed could be considered a pulse disturbance if the natural history of the system suggests consumers can go locally extinct for similar lengths of time. However, it is often unclear what constitutes a natural event or a different community because levels of natural variation in disturbance or species composition are rarely specified beforehand.

At times it is too difficult or unethical to manipulate an entire community, and so proxies are used instead. Proxies come in two types. In some cases a single species is used as a sentinel for an entire assemblage, and experiments use that species as an indicator of a community type. Redwood forests are defined by the presence of redwoods and mussel beds by the presence of mussels, and so it is plausible that the population dynamics of a sentinel species can define the rise or fall of the whole community. Since sentinel species are often ecosystem engineers, many of the other species are simply riding piggy-back on the sentinels.

Yet sentinel species can be extraordinarily long-lived, and so it may impossible to carry out an experiment over a long enough period to show changes in their population densities that are ecologically meaningful. In these cases, differences in life-history processes (e.g. recruitment rates of various species after a fire) are used as proxies for changes in per-capita rates of population growth. Relatively short-term experiments are then used to demonstrate the plausibility that changes in these processes could lead to different species assemblages.

This brings us to the second type of proxy, which involves using rates. Using rates of recruitment, mortality, or, in some cases, changes in body weight or size as proxies rely on three assumptions. First, we must assume that short-term estimates of the rates of individual-level processes are good proxies of per-capita rates of population growth. Second, these processes will continue at the same rate for long enough to drive the system into a different basin of attraction. Third, it is assumed the initial direction of the per-capita rate of population growth after a perturbation points directly into a basin of attraction. This assumption is naively embedded in the typical cup and ball analogy for stability and basins of attraction in which the ball rolls downhill and takes the shortest path to the bottom of the basin. Yet this is clearly not the case in even a simple Lotka–Volterra model of a predator–prey system with a stable focus. After a perturbation, prey and predator populations may appear to move away from the stable point as they begin to cycle. In addition, experimental results are at least two steps removed from the target of interest. Rates of individual processes are proxies for population dynamics of sentinel species, which in turn are proxies for changes in community structure and composition.

Finally the most troublesome issue is self-replacement, which is a proxy for stability (Connell 1983, Grimm and Wissel 1997, Peterson 1984, Sousa and Connell 1985, Sutherland 1990). The theory of multiple stable states was developed in the context of dynamical systems where stability is well-defined and the question of stability has either a yes or no answer (Lewontin 1969). In practice, ecologists must use a proxy, and self-replacement is just one of many (Grimm and Wissel 1997). This is the Achilles heel of all experimental tests of multiple stable states whether or not Peterson's criteria are used.

Grimm and Wissel (1997) suggest that ecologists stop trying to do the impossible and instead use commonly accepted properties of stability and define carefully the situations in which these properties are being used. Grimm and Wissel (1997) reviewed 163 definitions of stability that have been used by ecologists and grouped these definitions into six broad categories. These are persistence, constancy, resilience, resistance, elasticity, and domain of attraction. Most were anticipated by Lotka (1956) and Lewontin (1969). Both Lotka and Lewontin noted that constancy did not imply stability and that stability was a dynamical property. Lewontin (1969) also noted that presence of oscillatory behavior was important because weaker oscillations during the return to an equilibrium point could be viewed as "more stable." Grimm and Wissel also downplay the fact that even relatively simple models can produce extraordinary dynamics that include long transient times, fractal basins of attraction, and strange attractors (e.g. Grebogi et al. 1987, Huisman and Weissing 2001a, b, Savage et al. 2000).

Following Grimm and Wissel's terminology, persistence and constancy are properties of a reference state in the absence of extrinsic disturbances or perturbations. Persistence is the existence or endurance of a reference state or condition over time while constancy is the property of "staying essentially unchanged." It is possible to imagine an ecological condition being persistent but not constant, for example the population size of fleas on a dog, but it is hard to imagine constancy without persistence. The next three properties—resilience, resistance, and elasticity—are descriptions of how a system reacts in response to a perturbation. Resistance is the lack of response despite perturbations, and a number of authors have used either inertia (Murdoch 1970, Orians 1974) or sensitivity (Estberg and Patten 1976) to describe the same phenomenon. Resilience is the ability to return to a reference state after a disturbance, while elasticity is the speed of that return. Both resistance and resilience are defined in terms of the "size" of the perturbation; systems that are able to resist or return to initial conditions after a large perturbation show greater resistance or resilience. Finally, Grimm and Wissel define the domain of attraction as the "whole of states from which the reference state... can be reached... after a temporary disturbance." This definition appears to match Lewontin's (1969) concept of the size and extent of the basin of attraction, and perhaps redundant with the idea of resilience. This can be seen by remembering how perturbations in dynamical systems are metaphors for disturbances in ecological communities. Strictly speaking, perturbations in dynamical systems are changes in state variables (densities, biomass, etc) that are caused by extrinsic forces. Ecologists often use the term perturbation more broadly, and often include temporary changes in parameters (i.e. environmental conditions) that in turn shift the state variables. There state variables are altered indirectly via a parameter that is intrinsic to the system.

Our discussion of Peterson's criteria suggests that an appropriate experimental protocol should have seven aspects. For simplicity, assume we have only two alternative community states and, as in Figure 3.1, we are following the densities of a sentinel species in each community state. The same logic would apply if we were using a multivariate metric to track the entire assemblage, or following an ecological process associated with the sentinel species. First, an assemblage of species or a sentinel species that is at equilibrium needs to be identified (e.g. N^*_{1i} in Figure 3.1). Second, the assemblage or species must be perturbed so that the ridge line is crossed. Since is it rarely known where the ridge line exactly lies, a set of perturbations—"small to large"—should be done to bracket it. If there are multiple stable states, then "small" perturbations should fail to tip the system while "larger" ones do so. Third, the perturbation must be a pulse disturbance and should mimic a natural event in spatial extent, temporal duration, and its effect on species. It is important to recall that whether a perturbation is a pulse or a press is defined by the system and its natural history. Fourth, after the perturbation, the system cannot be manipulated and must be left to "run free." Fifth, the system must be followed for long enough so that an alternative state will develop. Sixth, the reverse experiment must also be done. If species 1 in the system is perturbed from current equilibrium point (N^*_{1i}) and observed to diverge to state j (N^*_{1j}), then reciprocal experiment should also be done. Species 1 in state j should be perturbed and observed to diverge to its density in state i. Ideally, the two experiments should be done in exactly the same place to meet Peterson's requirement. Practically, this will rarely be possible. What can be done is to carry out one experiment in places containing community state i and the reciprocal experiment places containing in state j. At best, this assumes that Peterson's very specific requirement of the "very same site" can be met by using a large number of replicates in the same "environment," yet it is a very real possibility that different communities actually occur in slightly different environments. Thus the final aspect is that both experiments need to be well replicated and the replicates from each experiment interspersed temporally and spatially to avoid simple and temporal pseudoreplication (Hurlbert 1984). Good replication will provide some protection against the problem of undetected environmental variation, but it is not a guarantee.

The presence of environmental variation means that identical perturbations may not give the same response every time or at every place. There are two ways in which this could occur. First, parameters could vary within the context of the system, and in terms of dynamical models, the variables do not change but the parameters may vary. Note that if some of the parameters can be zero or change sign, it would seem as if the model changed when in fact it has not. Lewontin (1969) described this sort of variation as changes in the vector field, and it is equivalent to his analogy of a ping pong ball in a bowl that is being constantly wiggled. The ball (the state variables) moves as the position of the bowl (the parameters) is changed.

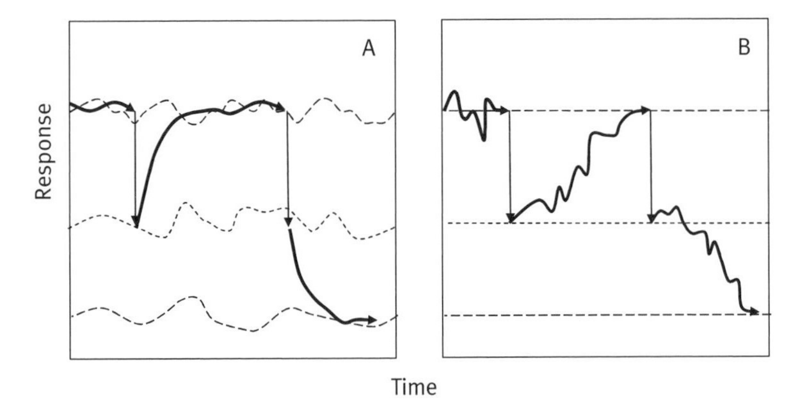

Figure 3.2 Responses of state variables to intrinsic and extrinsic changes. Panel A shows how changes in parameters will cause the position of the equilibrium points and the ridge line to change over time and how the state variables will try to track these changes. Panel B shows how extrinsic changes alter the state variables but do not affect the position of the equilibrium points or the ridge line. Note that in both panels the same perturbation can result in two different outcomes.

Equilibrium points will shift as parameters vary, and so over time we will observe haphazard changes in position of the equilibrium points and the ridge line (Figure 3.2A). Thus in one experimental replicate, a controlled pulse perturbation of densities may cross the ridge line, and in another replicate the same perturbation may fail to do so. Schröder et al. (2005) call this a test for random divergence and suggest it as a good test for multiple stable states. Interestingly, Schröder et al. (2005) discussed the phenomenon of random divergence in the context of hysteresis, even though it is possible to have divergence occur in linear systems with multiple stable states. The problem, however, is that identical perturbations can only be done either in different places at the same time or at different times in the same place. It is difficult if not impossible to prove that the environment is indeed the same at different places and at different times. Divergence may be nothing more than different responses in state variables to differences in parameters. Again, sufficient replication and randomization of replicates can help protect against this problem, but there is no way to be certain that in fact the replicate runs of an experiment were under the same environmental conditions.

In addition, if parameters vary, population densities are unlikely to be at equilibrium because the per-capita rates of change for species are likely to be slower than the rates at which parameters fluctuate. Thus species densities will always lag behind and the system will never reach a stable equilibrium point. If species compositions of a set of replicate runs are not close to an equilibrium point, we might interpret variation among replicates as random divergence, particularly if the set of replicate runs shows a slight multimodal pattern. This is, in essence, what underlies Hutchinson's (1961) resolution of the paradox of the plankton where species compositions never quite come to equilibrium, and so individual species are rarely driven to extinction by competitive exclusion.

It is also possible that densities change due to "extrinsic" perturbations and not as the result of changes in parameters. In Lewontin's analogy of a bowl and a ping pong ball, the bowl is now stationary and the ball itself being flicked around inside the bowl by an external force such as our fingers. The equilibrium densities and the position of the ridge line remain constant over time, but since extrinsic perturbations affect densities, the trajectory after a large disturbance does not follow a smooth exponential approach to either equilibrium point (Figure 3.2B). Identical disturbances will not give the same response, with some replicate runs appearing to return quite quickly to initial conditions and other replicate runs appearing to take a very long time to do so. In the extreme case, if small extrinsic perturbations push densities across the ridge line then identical disturbance will cause some replicates to return to the original state and others to diverge to the alternative state.

3.2 Mutual invasibility and priority effects as tests for multiple basins of attraction

A second approach for detecting multiple basins of attraction involves showing that one community cannot be invaded and taken over by members of another community even though both communities occur in the same environment. The classic example of this is the Lotka–Volterra system of two competitors with an unstable saddle node in which each competitor is unable to invade if the other is close to its own stable equilibrium point. There are many examples of mutual invasibility, but these are not usually considered as examples of multiple stable states even though they are. Interestingly, priority effects, in which the order of arrival of species determines the final species composition, are often cited as way for alternative stable states to arise, and yet priority effects are, in fact, nothing more than an extension of two-species mutual invasibility to a multiple-species context.

By modifying the definition of perturbation and using two sentinel species, the same protocol could be used with demonstrations of mutual invasibility as tests of multiple stable states. The perturbation in this case is the introduction of a sentinel species (species 1) from one community (state *i*) into an established alternative community (state *j*) and observing that the sentinel fails to invade. The reciprocal experiment would involve using the second sentinel species (species 2) which is common in community state *j* and introduce into state *i*. Since both established communities do not physically occupy exactly the same place, it is not possible to meet Peterson's requirement of the "very same site," and the reciprocal introductions would have to be well replicated over a wide area of similar habitat to protect against the possibility that undetected environmental differences are the underlying cause for the different communities.

3.3 Demonstration of hysteresis

Demonstration or detection of hysteresis is a less inclusive approach because, as we shall see in Chapter 6, it is possible to have multiple stable states without hysteresis. To make matters worse, it is usually assumed that tests for hysteresis only require showing that a community once perturbed fails to recovery to the original state (e.g. see Figure 2 in Schröder et al. 2005). It is often assumed that the dynamics of an ecosystem can be explained by hysteresis when three seemingly contradictory phenomena are observed. First, the system appears to be very resistant to range of environmental changes. This is often observed as little or only slight shifts in density of a sentinel species, which is a state variable, with incremental changes in environmental conditions (a model parameter). Second, a sentinel species unexpectedly undergoes a large shift in density with additional changes in environmental conditions, and the system is tipped to a different state. Finally reversal to the original environmental conditions (parameter values) will not return the system to the original community structure (i.e. the state variables).

However, simply demonstrating nonrecovery is not a complete and sufficient test for hysteresis, and definitive tests must include a demonstration that the community-level response is asymmetrical with respect to changes in a parameter. This can be seen by comparing what happens in a system that undergoes a phase shift as environmental conditions change versus a system that shows hysteresis as conditions change (Figure 3.3). Phase shifts are abrupt but smooth transitions in species composition as environmental conditions change, and the plot of a state variable (e.g. density of a sentinel species) against a critical parameter, related to environmental conditions, will be an S-shaped curve with a steep shoulder (Figure 3.3A). As environmental conditions change so that the parameter value passes the shoulder (from x to y), the density of the sentinel species crashes (Figure 3.3; see panels B and C). Reversing the environmental conditions (from y to x), returns the density of the sentinel species to initial levels.

In contrast, a system with hysteresis will not show recovery. The plot of density versus the critical parameter will have a two folds, and there will be an abrupt and discontinuous shift in equilibrium density at points a and b (Figure 3.3D). Density may change rapidly as point a is passed with changes in environmental conditions from x to y. However, reversing conditions (from y to x) will not cause the system to recover. Environmental changes must be sufficiently altered so that point b is passed, at which time the density of the sentinel species will recover. Typically this pattern of nonrecovery is taken as prima facie evidence for hysteresis.

The problem, however, is that density changes over time will look identical for systems that have smooth phase shifts but not hysteresis (Figure 3.3; compare panels C and F). Moreover, the S-shaped curve is a plot of equilibrium densities and may not reflect how fast transient dynamics will occur. It is also not possible to distinguish between the two unless environmental conditions are monitored to ensure identical forward and backward shifts in conditions. To make matters worse, time lags in the response of state variables can make systems with phase shifts appear not to recover.

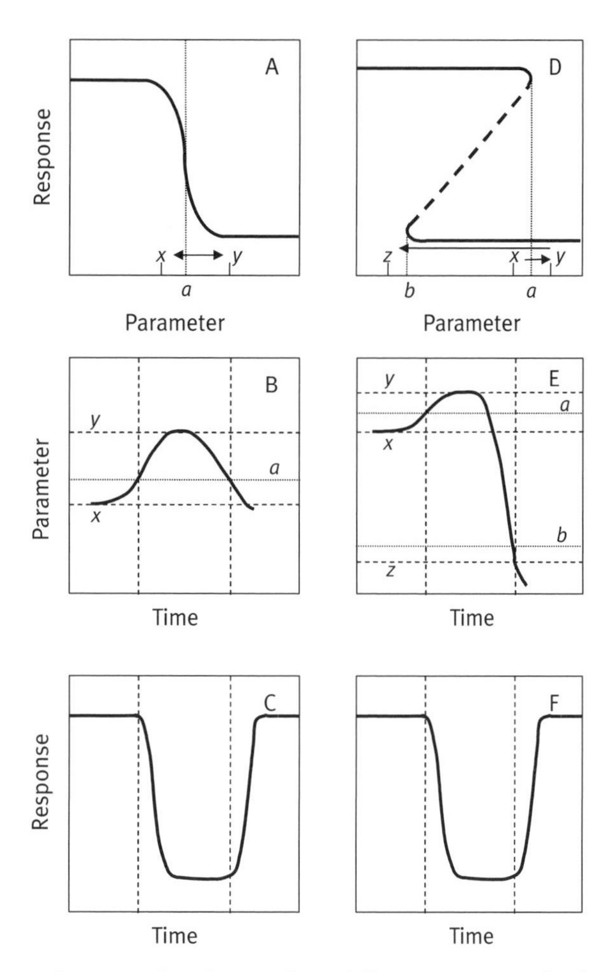

Figure 3.3 Responses of a system that shows a phase shift versus a system that has hysteresis. Panels A, B, and C show a system with smooth phase shift and panels D, E, and F a system with hysteresis and discontinuous jumps.

Systems with hysteresis are often said to show "resistance" or "lag behind" changes in environmental conditions, but the same is true of systems with time lags. This is a very serious problem, since many ecosystems that have been suggested to show hysteresis also often contain long-lived sentinel species whose responses to environmental change are likely to lag behind simply because of demographic considerations.

Compelling tests for hysteresis must then take in to account the possibility that what appears to be nonrecovery is not hysteresis. Paradoxically, experimental designs that could distinguish between hysteresis and time lags require the use of press rather than pulse perturbations. Imagine that environmental conditions are shifted experimentally from *x* to *y* (as in panels A and D in Figure 3.3) and held at *y* for an extended period of time. The change in environmental conditions crosses the shoulder in systems with

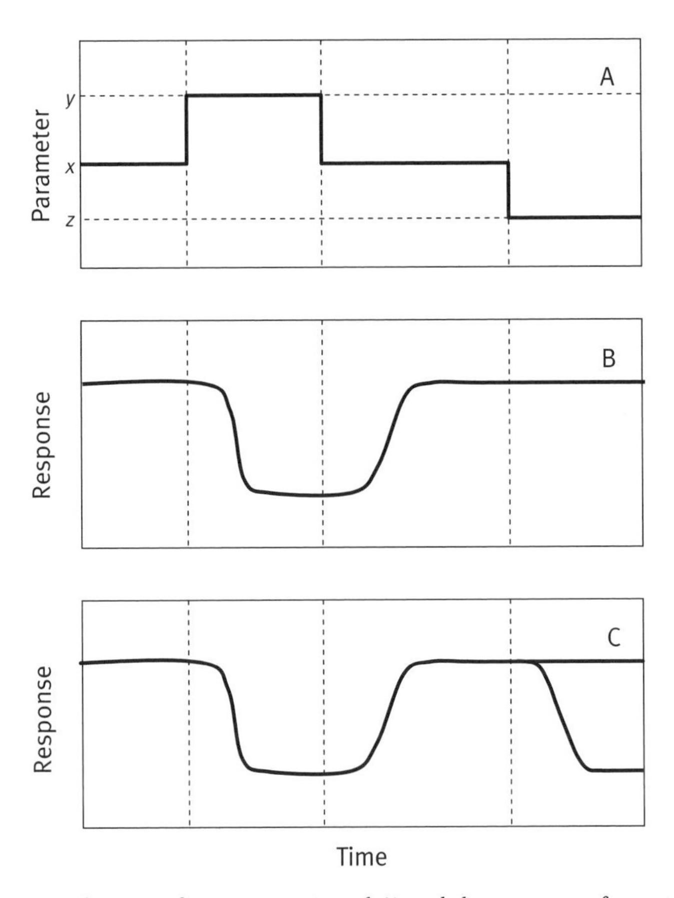

Figure 3.4 Press perturbations of parameters (panel A) and the response of a system with phase shifts (panel B) versus the response of a system with hysteresis (panel C).

phase shifts and the edge of the fold in systems with hysteresis, and in both systems there will be a dramatic shift in species composition at equilibrium. However, if there are time lags, the response in species composition may lag behind the shift in environmental conditions, and this will be true in both types of systems.

This can be seen in Figure 3.4 where the change in species composition lags behind the shift in environmental conditions in a system with a phase shift and in a system with hysteresis. The responses in the two systems could be identical, and distinguishing between them requires two additional steps. First, the two systems will respond differently to a reversal of environmental conditions (i.e. going from y to x). In a system with a phase shift, species composition will return to the initial state, but in a system with hysteresis, there will be no change. Second, the response in the two systems will also differ if environmental conditions are pushed back even more so that the opposite edge of the bifurcation fold is passed (i.e. going from x to z, see panels A and D in Figure 3.3). There will be no change in species composition in a system with a phase shift, but a

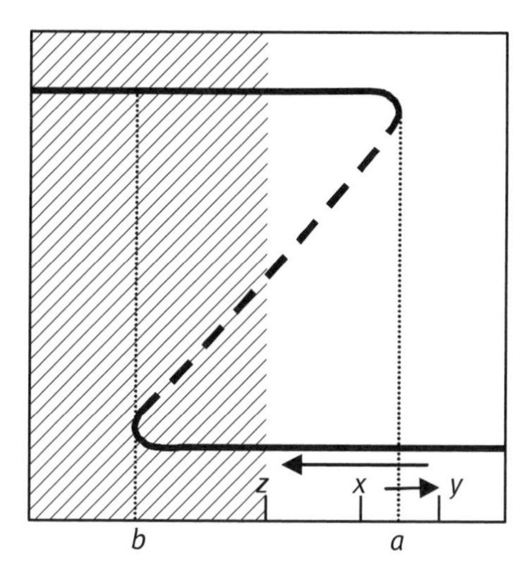

Figure 3.5 Hysteresis in an ecosystem with a truncated range of environmental conditions (shaded area). A change in the parameter from *x* to *y* changes a discontinuous jump in the equilibrium state. A reversal of change in the parameter can go only as far as *z*, which is the minimum value seen in the field. The system can jump back the original state because *b* does not exist in the environment.

return to initial species composition in a system with hysteresis (Figure 3.4). All three manipulations (*x* to *y*, *y* to *x*, and *x* to *z*) must be done with press perturbations to control for the effects of time lag and to allow enough time for the system to respond.

It is also possible that a system may have the potential for hysteresis but it is never seen in nature because the range of environmental conditions is truncated and does not include both folds of the S-shaped curve (Figure 3.5). Thus the system could be moved from one to another state (*a* to *b*) with changes in environmental conditions (*x* to *y*) but the original state could not be recovered because the range of environmental conditions is not broad enough. In this sort of system it would appear as if the change in community state was irreversible.

All the caveats about the use of proxies for state variables and the effects and sources of variability that were discussed in relation to testing Peterson's criteria apply to the detection of hysteresis. Using proxies such as measures of ecological processes as proxies for state variables—for example using rates of recruitment or mortality as an index of the presence of a species—assumes there is a one-to-one correspondence between a process and a state variable. Often this will not be true. Using sentinel species is even more problematical because they tend to be very long-lived species, and so it is almost certain that sentinels will show a lag in response to changes in environmental conditions (i.e. parameters). Thus it will be very difficult to distinguish between a system with time lags and one that truly has hysteresis.

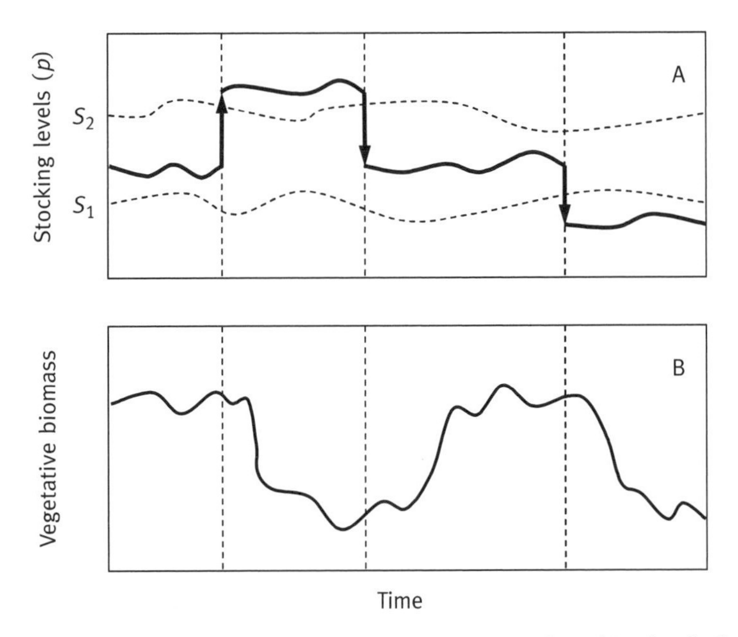

Figure 3.6 Sources of variation. Panel A shows variation in critical stocking levels that is due to changes in parameters (dotted lines) and in experimental control of press perturbations of stocking levels (solid lines). Panel B shows response of biomass to these sources of variation plus added variation caused by extrinsic disturbances to biomass.

Sources of environmental variation present problems because demonstrations of hysteresis involve using press perturbations to change parameter conditions. Thus there may be variation not only in state variables and parameters but also in experimental control of the press perturbation. The three types of variation can be easily distinguished by re-examining May's (1977) model of grazing by range animals (Equation 2.10). Recall that the rate of growth of vegetation is $g(N)$ where N is vegetative biomass and the rate of removal by grazers is $ph(N)$ where p is the number of grazers and $h(N)$ is the per-capita rate of consumption. Environmental variation can affect the parameters that define $g(N)$ and $h(N)$ and this variation will cause the critical thresholds for stocking levels to vary over time (Figure 3.6). In addition, levels for vegetative biomass may be pushed off the equilibrium values by environmental conditions extrinsic to the model—for example, density-independent mortality caused by drought—and we would expect to see biomass vary over time even though parameters remain constant.

We have already mentioned these two sources of variation in our discussion of Peterson's criteria, but in contrast, the third sort of variation—variation in experimental control of the press perturbation—is new and a bit different. Recall that demonstrations of hysteresis require using a press perturbation of a parameter in contrast to tests of Peterson's criteria, which require pulse perturbations of state variables. It may be very difficult to hold a press perturbation—such as the stocking levels of grazers on rangelands—at a constant level for a long period of time, and so we might expect the perturbation itself to show variation over time (Figure 3.6). Taking all three sources of variation into account means that we should not expect to get the same result from doing the same experiment in slightly different places and at different times.

4

Experimental evidence

There have been very few good experimental tests of multiple stable states in natural ecosystems because the litmus test for multiple stable states in natural ecosystems is extraordinarily stringent. At first glance, Peterson's three criteria—same site, different communities, self-replacement—and the use of natural pulse perturbations appear to be unambiguous. However, they have proven difficult to meet in field experiments (Connell and Sousa 1983, Petraitis and Dudgeon 2004, Schröder et al. 2005, Suding et al. 2004).

Schröder et al. (2005) reviewed 35 experiments published between 1980 and 2004 for evidence of multiple stable states and concluded that only nine were good enough or appropriate to be considered field tests of multiple stable states (Table 4.1). Only 19 of the 35 experiments were done in the field, and of the nine good tests, Schröder et al. judged three to be supportive of the existence of multiple stable states. Even so, a closer look at these nine cases and several others suggests that very few experiments contain adequate controls and are well-replicated without pseudoreplication.

There are a large number of experiments that have been done in freshwater systems that are not cited by Schröder et al. (2005) yet are often mentioned as good tests for multiple stable states. In fact many involve press perturbations and do not meet the criteria outlined in Chapter 3. Most use microcosms or mesocosms and involve manipulation of either nutrients, which should be considered as parameters, or fishes, which could be either a state variable or a parameter (e.g. Irfanullah and Moss 2004, Irvine et al. 1989, Jones et al. 2002, Morris et al. 2003a, b, Portielje and Roijackers 1995, Potthoff et al. 2008, Thomas and Daldorph 1994). Many lack adequate controls and are pseudo-replicated. Press perturbations are de facto changes in environmental conditions (i.e. parameters), and the demonstration of two different states under different environmental conditions is non-informative with respect to multiple stable states. In nutrient studies, phosphorus input is usually experimentally manipulated and responses of macrophytes and/or algae are followed over time (e.g. Bakker et al. 2010, Irfanullah and Moss 2004, Jones et al. 2002, Thomas and Daldorph 1994; but see Stephen et al. 1998 for manipulation of nitrogen levels). In some cases, natural variation in nutrient levels over several ponds is used (Portielje and Roijackers 1995), although that particular study suffered from pseudoreplication. Experiments addressing top-down processes use similar protocols but in which fish rather than nutrients are manipulated (Irvine et al. 1989, Potthoff et al. 2008). Other studies have included factorial designs of nutrients and predators (Stephen et al. 1998).

Table 4.1 The nine experiments cited by Schröder et al. (2005) in their Table 1A as good tests of multiple stable states. Their literature search spans from 1980 to 2004.

System	Pro/ Con	Comments	*N*	Pseudo-replicated?	Reference
Marine					
Fouling surfaces	–	Priority effects with persistent communities over the short term, but not self-replicating (+)	53[a]	Yes	Sutherland (1974, 1981)
Rocky intertidal	–	Return to initial state after pulse removal of seastars (–)	2[b]	Yes	Paine et al. (1985)
Rocky subtidal	–	Priority effects seen (+)	12–36[c]	Doubtful	Kennelly (1987)
Freshwater					
Small lakes	–	Return to initial state after pulse removal of fishes (–); a meta-analysis, not an experiment	18[d]	Yes	Meijer et al. (1999)
Artificial ponds	+	Divergence; see text for comments (+)	48	No	Chase (2003a)
Lake mesocosms	–	Failure to move the system to alternative state after a pulse removal of submerged vegetation (–)	9	No	Morris et al. (2003b)
Terrestrial					
Salt marsh	+	Press experiment, not a multiple stable state experiment (–)	16[e]	Yes	Handa et al. (2002)

Grassland	–	Convergence after cessation of grazing (–)	4[f]	Yes	Valone et al. (2002)
Old fields	+	Divergence if perturbed above a threshold (+)	60[g]	No	Schmitz (2004)

Pro/Con, Schröder et al.'s assessment of whether the experiment provided evidence pro (+) or con (–) for multiple states; Comments, our assessment; *N*, the total number of experimental units, which in some cases is the correct total number of replicates per treatment level; Pseudo-replicated, assessment of pseudo-replication based on Hurlbert's (1984) definitions.

[a] Total number of fouling plates were submerged and then followed for 6.75–7.75 years. Pseudo-replicated because treatment effect of timing of submergence on plates and repeated sampling of same plate are ignored.

[b] One plot in which seastars (*Heliaster helianthus* in Chile) were removed and one control plot in which they were not. Temporal pseudo-replication.

[c] Number of replicates per treatment cell is not clear because description in text does not match degrees of freedom (d.f.) given in analysis of variance (ANOVA) table. Pseudo-replication is unlikely but cannot be evaluated because of unclear description.

[d] Before–after design with removal of fishes in 18 lakes. Not correctly analyzed either as a meta-analysis or as an intervention analysis.

[e] Eight sites that differed in composition, One 5 m × 5 m exclusion per site; 15 subsamples inside and outside exclosures. The d.f. of ANOVAs are based on subsamples.

[f] Two sites and one fenced area in each site; 12 subsample transects inside and outside. Simple pseudo-replication.

[g] Description of ANOVA design (six treatment levels and ten replicates per level) does not match d.f. of ANOVAs (d.f. = 5 and 49).

All interpret—either explicitly or implicitly—the difficulties in returning a lake in the turbid state to a clear-water state as evidence for hysteresis. Almost all experimental studies fail to show what would happen under free-running conditions after a pulse perturbation. One exception is the study by Persson et al. (2007) in which fishes that are apex predators were culled from lakes. Culling of the apex predator allowed the prey (i.e. other fish species) to recover, and even with the return of the apex predator some lakes remained in an alternative state for 15 years. The re-invasion of predators could, in this case, be viewed as a very long pulse perturbation. Overall, it is not surprising that Barker et al. (2008) concluded that multiple stable states have rarely been demonstrated experimentally in lakes.

Schröder et al. (2005) also mention several studies that do not make it into their list of 35, and a number involve using proxies or demonstrating aspects of a system that are consistent with the existence of multiple stable states. Two well-known examples that have been widely cited as evidence for multiple stable state are the work of Augustine et al. (1998) and Konar and Estes (2003); both have been cited over 70 times. Augustine et al. (1998) undertook a grazing experiment involving white-tailed deer (*Odocoileus virginianus*) on a dominant understory forb (*Laportea canadensis*) in southeastern Minnesota. They set their experiments in context of Noy-Meir's models, and by using fences to keep deer out they effectively "tuned" the grazing parameter. The followed the areas for over 2 years and saw changes consistent with multiple stable states, but the evidence was indirect at best. The experiment also had very low replication (two to four replicate enclosures per experiment and only three experiments).

Konar and Estes (2003) addressed abrupt spatial shifts in their study of the boundary between kelps and barrens. They manipulated both kelps and urchins that are found in barrens. They provide good indirect evidence that the presence of kelps keeps out urchins and thus sets the boundary. They suggest the sharp boundary is the edge between the alternative stable states of kelps and barrens. They used proxies and did not follow the system long enough to go through even one turnover. Konar and Estes are also unclear if they are pushing the system across the breakpoint ridge or over a sudden jump. Manipulation of kelps involves changing a state variable, while manipulation of sea urchins, which are grazers, could be viewed as changing a parameter.

There are numerous examples of priority effects as multiple stable states. Priority effects involve showing that the final community state depends on the order in which a system is perturbed. This can be the order in which species are introduced or the timing of an abiotic disturbance. Many of the papers place tests of priority effects either in the context of the Lotka–Volterra model of two-species competition or in terms of community assembly rules (for discussion of community assembly rules see Drake 1990, 1991 and Drake et al. 1993). Cole (1983) showed that the order of introduction of two species of ants matters; the first species introduced and established always prevents invasion of the other. Schröder et al. (2005), interestingly, considered Cole's study inappropriate as a test of multiple stable states. Similar patterns have been found for protist communities manipulated in the laboratory (Jeppesen et al. 2007) and cladoceran zooplankton in experimental ponds (Louette and De Meester 2007). Along the same lines, it is known that the course of succession can depend on the timing of disturbance (e.g. Foster et al. 2003) and on the sequence order of disturbance (e.g. Fukami 2001).

The value of using priority effects as a test for multiple stable states is limited because of the same-site requirement (see Section 3.2). The requirement is easily met if experiments can be done in the laboratory or in experimental units such as microcosms in which replicate ecosystems can be established. This is more difficult if not impossible for field experiments. As we discussed in Chapter 3, there must first be an established community, and then we must show not only that an invading species can become established but also that the invader causes the creation of a uniquely different community state. We then must show that, if we have that new state, we cannot reverse the process. It would be extremely difficult to conduct a well-replicated experiment with adequate controls in the field without using artificial units such as experimental ponds.

Given the uncertainty of classifying studies as being supportive or not of multiple stable states, we will take a closer look at three experimental studies as exemplars of what is needed to fulfill Peterson's criteria. The first is Morley's (1966) comments on an earlier study of the effect of stocking density on weight of ewes. Morley's paper is cited by May (1977), and Noy-Meir (1975) suggested that Morley's comments about the sharp and bimodal distribution of the weights of ewes is the first experimental test of multiple stable states. Morley did not mention multiple stable states directly. The second is Chase's (2003a) well-known microcosm experiment of how the initial densities of freshwater prey can affect the final abundances of an insect predator and two of its prey, algae and macrophytes. Chase's experiment is one of the best, if not the best, microcosm experiment, and is one of the three experiments mentioned by Schröder et al. (2005) as providing evidence in support of multiple stable states. The third example is Petraitis et al.'s field experiment on mussels and seaweeds as alternative states (Petraitis and Dudgeon 1999, 2005, Petraitis et al. 2009). This large-scale study is one of the best replicated field experiment specifically designed to test for multiple stable states.

4.1 Morley's comments on live weights of ewes

Morley (1966) was interested in the factors affecting the joint stability of livestock production and pasture condition. Pasture stability for Morley was then the management target of maintaining high and consistent levels of livestock production regardless of changing environmental conditions. Production was measured as livestock body weights and either as the average per-capita weight or the average per-acre weight. Morley listed four categories of factors that influence pasture stability: animal physiology, plant physiology, ecological factors, and site-specific features (see his Table 5). Each category included a number of specific characteristics—for example, animal physiology included growth, reproduction, and maintenance requirements, and ecological factors included longevity and species composition of the plants present. Morley also noted that factors could be stabilizing or disruptive. Translating his terminology into more modern language, he suggested, for example, that a diverse mixture of generalist perennials tended to stabilize pastures while a monoculture of specialized annuals destabilizes pastures. This is a surprisingly modern idea and not very different from the present-day discussions of the links between biodiversity and the stability of ecosystem function. Morley also provided a graphical representation of the balance between plant growth and grazing, which he called animal intake (his Figure 2). Morley's graph is remarkably similar to figures in May (1977; his Figure 1) and Noy-Meir (1975; his Figure 5).

Morley noted that the average live weights of Merino ewes were either bimodal or unimodal and the switch between the two distributions depended on stocking densities (Figure 4.1). The average weight of ewes held at five animals per acre showed a unimodal distribution while those held at six and seven animals per acre had bimodal distributions. The bimodal pattern of live weights at six and seven animals per acre suggests there are two stable points at those densities. Thus as stocking density is increased from five to six and then to seven, the average weight declines with a sudden and discontinuous jump. Lowering the stocking density will not cause an increase in average weights until the system is pushed back below six animals per acre. The system shows hysteresis. While Morley did not use terms such as jumps or hysteresis, he did suggest that animal production may decline rapidly when a critical level of stocking was reached. He called these "crashpoints" or "thresholds" and placed both terms in quotes.

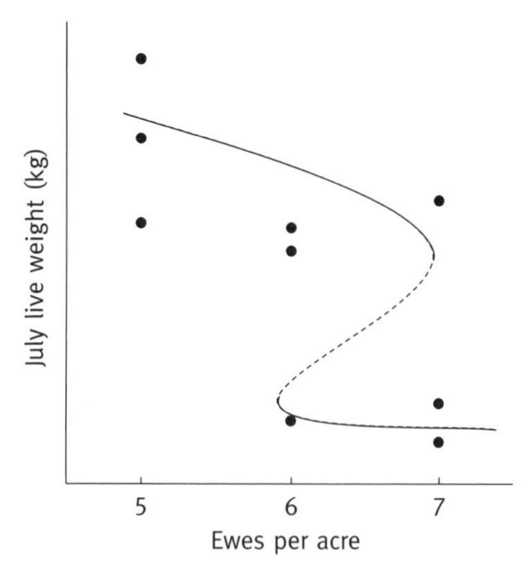

Figure 4.1 Morley's figure of live weights of ewes. Note that Morley did not provide a scale for weights in his original figure. Redrawn from Morley, F. H. W. (1966). Stability and productivity of pastures. *Proceedings of the New Zealand Society of Animal Production* 26, 8–21. Reproduced with permission from the New Zealand Society of Animal Production.

Morley gave enough information about his analyses so that we can look in detail at the changes in variation due to differences in stocking density. Morley analyzed the data from each stocking density and used a Model II analysis of variance to examine the differences among plots in average weights at five, six, and seven ewes per acre (see his Table 4). We can easily calculate the amount of variance due to variation among plots using the reported F-ratios and degrees of freedom along with the averages per plot (Figure 4.1). If the patterns of the plot averages were bimodal at six and seven ewes per acre and unimodal at five per acre, then we would expect to find larger variance components at greater densities. In fact, this is exactly what we see. Variation among plots as a percentage of the total variance accounts for 28% at five, 56% at six, and 74% at seven ewes per acre.

Morley ends his discussion of the experimental results with a number of very insightful observations that are strikingly similar to comments that can be found in the current literature (e.g. Carpenter et al. 2008, Scheffer et al. 2009). It is worth citing Morley in full:

> *First, an average response is a concept which... has very limited value in a particular situation.... Secondly, any farm, or paddock on a farm, may be subject to grazing pressure such that it hovers on the brink of a crash, without any recognized symptoms of danger being clearly observable.... These difficulties suggest that the true problem facing farmers and their advisers is the recognition of impending danger. The ecology of 'crashes' needs to be known so that they can be guarded against, and the causes rather than the symptoms treated. (Morley 1966, p. 12)*

The experiment discussed by Morley is not without problems. Morley reported the results of two experiments, each with two types of stocking. Only one of the four showed any evidence of multiple stable states. The two types of stocking were continuous in the same plot and rotating among three different plots in which animals were moved between plots. The same experiment with three stocking densities and two types of stocking was run in the summer and again in the winter (i.e. November and July in New Zealand). Only the continuous stocking treatment in the winter experiment showed any evidence of alternative states. There is also a problem with the stocking densities. Each plot held 12 ewes and so stocking densities were varied by changing the size of the plots. Densities of five, six, and seven ewes per acre means the plots were approximately 99 m × 99 m (2.4 acres), 90 m × 90 m (2 acres), and 83 m × 83 m (1.7 acres), respectively. Could the variation among plots in the live weights be related to the size of the plots? Pasture conditions often varies from place to place and on the scale of 10s of meters, so it is possible the slighter larger plots averaged out some of the spatial variation in pasture condition. If this were the case, then idiosyncratic differences among small plots could have contributed to the greater variation in average weights at higher densities.

4.2 Chase's experiment

Jonathan Chase (2003a) provides one of the most compelling tests of alternative stable states using microcosms, and it is worth taking a close look at his experiment. He showed that in a simple system with two species of freshwater snails as prey and an insect species that was a predator on both snail species it is possible to get two different outcomes depending on the starting densities of the prey species. He also manipulated the resources available to the prey species—that is the productivity of the system—and showed that alternative states only occurred at intermediate levels of productivity. Chase went to great lengths to place his experimental results within the context of a specific model (Chase 1999), which is a variation on a model proposed by Holt et al. (1994) and which we discussed in Chapter 2. Chase's experiment tested whether a predator and two-prey system could contain multiple stable states and if the occurrence of different states depended on the level of productivity. Models of one predator, two prey, and one resource predict that multiple stable states. are possible at intermediate levels of resource while at low and high levels only one of the two prey would coexist with the predator (Armstrong 1979, Holt et al. 1994, Chase 1999; see Section 2.3).

The two prey species, *Helisoma trivolvis* and *Physella gyrina*, are species of pulmonate snails that are commonly found in ponds throughout North America. Both snails are grazers and feed not only on algae, vascular plants, and periphyton, but also detritus. Both reach sexual maturity within 1 to 2 months, and females of both species can lay 10 to 100 eggs per week. *Helisoma trivolvis*, the ramshorn snail, can trap air in its shell and often occurs on floating vegetation, and it is also used in aquaria to control algal growth. As a result of its ability to float and its use by hobbyists, *H. trivolvis* has been accidentally introduced into ponds worldwide.

The predator, *Belostoma flumineum*, is the common giant waterbug (Hemiptera) and is found in western North America. The species has large forelegs and a strong beak, which allow it to hold and pierce it prey; they feed via their beak by secreting digestive enzymes into the prey and then drinking the liquefied tissue. Waterbugs are voracious sit-and-wait predators feeding not only on snails but also on crustaceans and amphibians. Large *P. gyrina* can escape predation by *B. flumineum*; *H. trivolvis* does not have a refuge in size (Chase 2003a).

The standing biomass of prey in natural ponds, which is correlated with productivity, is highly variable, and particularly so in ponds with intermediate levels of productivity (Chase 2003c). Moreover Chase found that natural ponds at intermediate productivity tended to fall into two classes depending on the biomasses of predators and large prey, which could escape predation. Ponds either had very large or very small biomasses of predators and large prey., The differences in biomasses were 2.5-fold for predators and 48-fold for large prey. There were only slight differences in pond area, productivity, total nitrogen, total phosphorus, and biomass of small prey.

Chase used 380-L tanks as his microcosms and manipulated primary productivity and initial densities of prey and predator. Tanks were stocked with phytoplankton, periphytic algae, vascular macrophytes, and zooplankton, and primary productivity was manipulated by the addition of phosphorus (NaH_2PO_4) and nitrogen ($NaNO_3$) to give three levels of productivity (low, intermediate, and high). The initial concentrations of phosphorus and nitrogen were within the range of natural variation. The two prey species—*P. gyrina* and *H. trivolvis*—were added to create low or high starting densities. *Belostoma flumineum* nymphs were added as predators to half of the tanks 1 week later. Chase then followed changes in the tanks for 180 days.

When predators were present, two alternative stable states developed at intermediate levels of nutrient enrichment depending upon the starting densities of prey. There is a clear S-shaped curve (Figure 4.2). In tanks that were initially stocked with high densities of prey, the final stable state had a large biomass of prey and a low biomass of producers. In contrast, tanks with low initial densities of prey stabilized into an alternative state with fewer prey and more producers. At the end of the experiment, the standing biomass of prey in tanks that started with high densities was over five times greater than the standing biomass of prey in tanks that started with low densities. There were also shifts in composition. Tanks starting with high densities of prey ended up with *H. trivolvis* as most of the prey biomass (98%) and algae as most of the producer biomass (89%). Tanks with low initial densities contained mostly *P. gyrina* (73% of the biomass) and a mixture of algae and macrophytes (66% of the biomass was algal).

Alternative states did not develop if predators were absent. The biomass of prey and producers increased with increases in nutrient enrichment. Tanks contained mixtures of the two prey species and of algae and macrophytes. There were no obvious changes in species composition across the enrichment gradient except for dominance by algae in tanks that underwent low enrichment. In the absence of predators, Chase found only slight differences in the effects of starting conditions on final biomass and composition at both low and high nutrient enrichment.

4.3 Rocky intertidal shores on the Gulf of Maine

Much of the rocky intertidal shore in protected bays throughout the Gulf of Maine is a mosaic of mussel beds and rockweed stands. Mussels are a common filter-feeding bivalve that often lives attached to solid surfaces but can also form large continuous beds on mudflats[1]. Rockweeds are large brown algae, and the two common species are *Fucus vesiculosus* and *Ascophllyum nodosum*. *Fucus vesiculosus* is a leafy species that may reach 50 cm in length, while *A. nodosum* is a rope-like species that varies in length but often reaches 1 m and occasionally exceeds 2 m. *Fucus vesiculosus* and *A. nodosum* have a forest-like appearance at high tide as the fronds float up into the water column and form dense mats on the surface at low tide. Mussels and rockweeds are ecological

[1] There are two morphologically indistinguishable species of mussels in the Gulf of Maine: *Mytilus edulis* is much more common in the south and *Mytilus trossulus* is more common in the north.

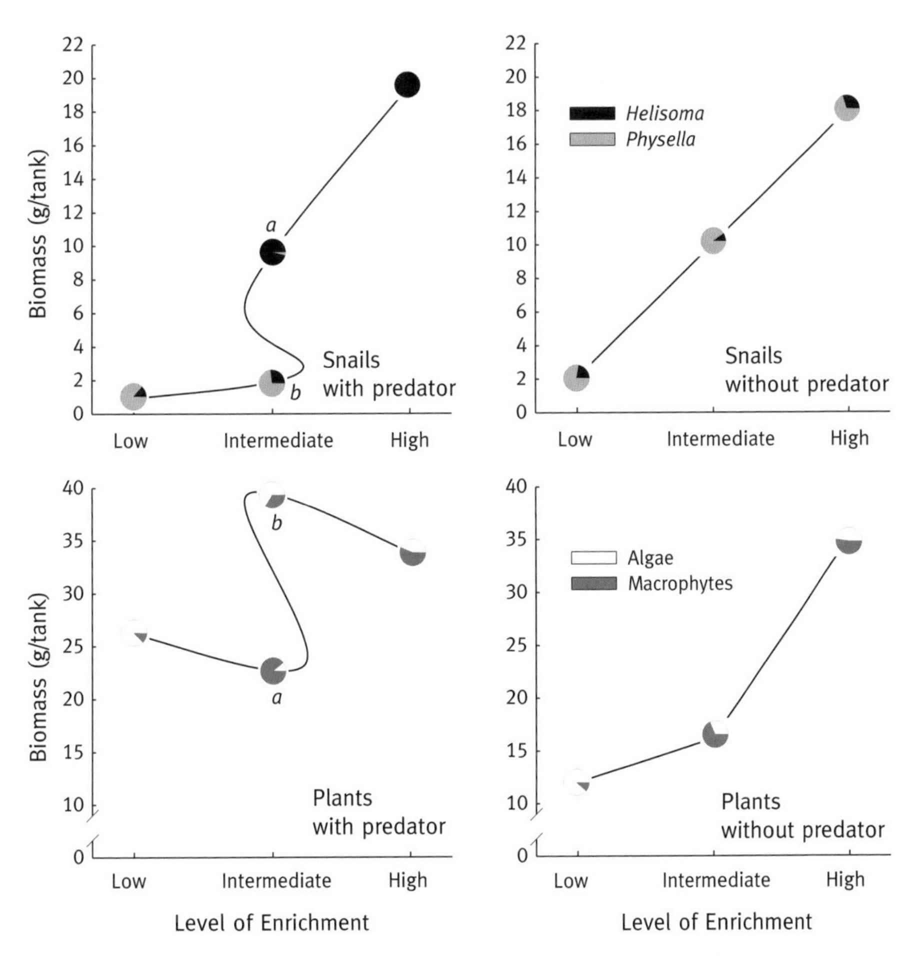

Figure 4.2 Alternative stable states of prey and producers at intermediate levels of nutrient enrichment in Chase's (2003a) experiment. Panels give final biomass and composition of prey species and producers with and without predators and at low, intermediate, and high levels of nutrient enrichment. Alternative states only occur at intermediate levels of nutrient enrichment and in the presence of the predator and depend on starting conditions for prey. Points a and b identify the two states at intermediate levels of enrichment. For example, the pair of points a shows the state with relatively high biomass of snails (upper right panel) and relatively low levels of plants (lower right panel). The data points here are based on averages calculated from the points provided by Chase (2003a, his Figure 3). Lines are fitted by eye, and the S-shaped curves are inferred.

engineers in that they alter the local environmental conditions. Many other species live in and use the structure provided by the presence of mussels and rockweeds.

It is commonly thought that the distributions of mussels and rockweeds are controlled by a combination of competitive interactions and predation. In a series of well-known experiments, Bruce Menge and Jane Lubchenco showed that mussels can out-compete rockweeds for the occupancy of rock surfaces and rockweeds tend to be common in areas where mussels are eliminated by predators (Lubchenco and Menge 1978, Menge 1976). The ability of predators to consume mussels is hampered by wave surge, and so mussels escape predation on exposed shores and can thus dominate the surface. In quieter waters, predators are more effective and reduce the number of mussels, and rockweeds are able to flourish in these areas.

Menge and Lubchenco's experiments fail to explain the co-occurrence of mussels and rockweeds in bays that are sheltered from wave surge, and Petraitis and Latham (1999) hypothesized that the mosaic of rockweed stands and mussel beds could be an example of alternative stable states. They saw the system as a balance between within-patch dynamics that are resilient against small perturbations, and thus maintain the status quo, and large-scale disturbances that were capable of flipping the system from mussels to rockweeds or vice versa. Petraitis and Latham suggested that ice scour, which is common in protected bays in the northern part of the Gulf of Maine, can clear large areas and could be the driver that provides the opportunity to tip the system from rockweeds to mussels. Ice scour can clear areas of the order of hundreds of square meters. At this spatial scale, the dynamics within a clearing become uncoupled from the surrounding environment and it is then possible for the system to be pushed into a different basin of attraction.

Petraitis, Dudgeon, and their colleagues (Petraitis et al. 2009) undertook a novel approach to test for multiple community states on Swan's Island, which lies off the coast of Maine. They first created experimental clearings in *A. nodosum* stands to mimic the effects of ice scour. Rockweeds are more common than mussels, but commonness tells us much about the competitive interactions (Box 4.1). The clearings were 1, 2, 4, and 8 m in diameter, and the larger sized clearings were close to the average size of clearings created by ice scour. They also included control plots that were not cleared. The successional changes in these clearings were then followed for 9 years. Next, observational data from persistent rockweed stands and mussel beds throughout the Gulf of Maine were collected to provide a priori definitions of different community types. These data were used to define benchmarks for the different community states that were independent from the data collected from the experimental clearings. Finally they used the benchmark data to assign each experimental plot to a unique community state and to assess ecological persistence. They only focused on the switch from rockweeds to mussels and not the reverse switch from mussels to rockweeds because the traditional view is that rockweeds persist only in areas where predators control mussels. According to this view, mussel beds should not occur on sheltered shores where predators are common and active (Bertness et al. 2004b, Lubchenco and Menge 1978, Menge 1976). Thus this is a strong test of multiple stable states because mussels are not expected to successful.

Petraitis and his colleagues discovered that the experimental plots could be assigned to very distinct groups based on clearing size. Small clearings, those which were 1 m in diameter, tended to return to being dominated by *A. nodosum* although a few (about 4.3%)

Box 4.1 Commonness tells us little about underlying dynamics

Rockweed stands are far more extensive and more common than mussel beds in protected bays, and it has been suggested that this is prima facie evidence for mussels being limited by predation. However, it is just as likely that the amount of cover by rockweeds versus mussels is the result of differences in the longevity of rockweed stands versus mussel beds (Petraitis et al. 2009). Individual mussels likely live less than 5 years and the persistence of a mussel bed depends on consistent recruitment of new individuals into the bed. It is likely that mussel beds on rocky shores and embedded in rockweed stand last for no more than 20 years. In contrast, individual *A. nodosum* plants may live more than 100 years (Åberg 1992a, b), and it is possible that rockweed stand could persist for hundreds of years. If stands of *A. nodosum* hold patches of rock at least ten times longer than mussels and if both species have the same chance of colonizing a newly opened patch, we would expect *A. nodosum* to be at least ten times more common than mussels (Petraitis et al. 2009). Longer-lived species will appear more common. A similar line of reasoning has been used to point out that prey items that take longer to handle and digest will be more common in gut contents, and as a result we may incorrectly infer that these prey items are preferred by predators (Fairweather and Underwood 1983, Peterson and Bradley 1978, Petraitis 1990). Commonness or the lack thereof tells us nothing about the underlying dynamics.

switched to mussel beds. For the clearings that were 2 m or more in diameter, 38% became mussel beds and the rest were dominated by *F. vesiculosus* stands (Figure 4.3). In a separate experiment, they showed that the median time of patch survival was between 3.1 and 6.4 years (95% confidence limits for smaller estimate 2.3–4.2 years, and for larger estimate 5.2–8.0 years; P. S. Petraitis and S. R. Dudgeon, unpublished data).

The results of Petraitis et al. show a clear bimodal pattern for both mussels and rockweeds in large clearings. Some clearings have about 40% mussel cover on the underlying rock and about 20% canopy cover by rockweeds while other clearings have fewer than 20% mussels and nearly 90% rockweeds. Control plots, which were left untouched for the entire 9 years, had few mussels (<10%) and an abundance of rockweed (>80%). The development of both mussel beds and rockweed stands shows how slight differences in parameters and state variables can give rise to both communities. This divergence is one of the characteristics of multiple stable states and we will return to divergences when we place the current notions of ecologists about multiple stable states in the more general context of catastrophe theory in Chapters 5 and 6.

It is quite striking that Petraitis et al. were able to induce the formation of mussel beds in experimental clearings and to show persistence of mussel beds without the use of predator exclusion cages in areas normally dominated by rockweeds. Previous researchers concluded that rockweeds could only persist if mussels were controlled

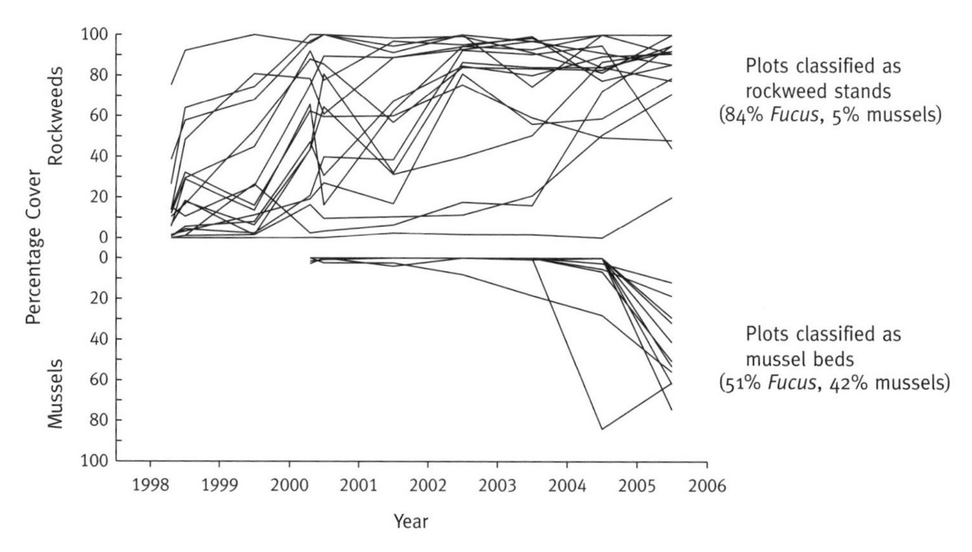

Figure 4.3 Divergence of large clearings towards alternative states of rockweed stands and mussel beds. Only clearings greater than 2 m in diameter are included; several data points were imputed because of missing data. Assignments were based on analysis in Petraitis et al. (2009); one of the 17 tracks diverging towards dominance by *Fucus vesiculosus* was classified as an *Ascophyllum nodosum* stand rather than a *F. vesiculosus* stand. Two of the tracks diverging towards mussels were classified as *F. vesiculosus* stands even though mussels were >50% under the canopy of *F. vesiculosus*. Based on unpublished data (Petraitis and Dudgeon).

by predators (Bertness et al. 2002, Lubchenco and Menge 1978, Menge 1976). Since protected bays are dominated by rockweeds, it was assumed that predators of mussels must be actively controlling their success. The fact that mussels can become established and persist if the system is tipped far enough strongly suggests mussel beds and rockweed stands are alternative states in protected bays.

Plotting Petraitis et al.'s results together with Menge's results suggests that mussels and rockweed are alternative states in protected bays (Figure 4.4). Menge's results show no indication of two states. The two sites exposed to the most wave surge—Pemaquid Point and Chamberlain—are dominated by mussels, but at Grindstone Neck, Menge's most sheltered site, mussels cover less than 20% of the rocky shore. Pemaquid Point is very wave swept, and so it is not surprising that it has almost no rockweed canopy, it being routinely torn off by waves. Rockweeds, and in particular *F. vesiculosus*, flourish at Chamberlain and Grindstone Neck, and cover 60 to 80% of the surface. Petraitis et al. (2009) used Grindstone Neck as one of their benchmark sites; their data, which were collected in 2005, are in close agreement with Menge's data which were collected in the mid 1970s (compare points 3 and 4 in Figure 4.4). In addition, the experimental clearings that were classified as rockweed stands are also very similar to Menge's data (compare points 3 and 6 in Figure 4.4).

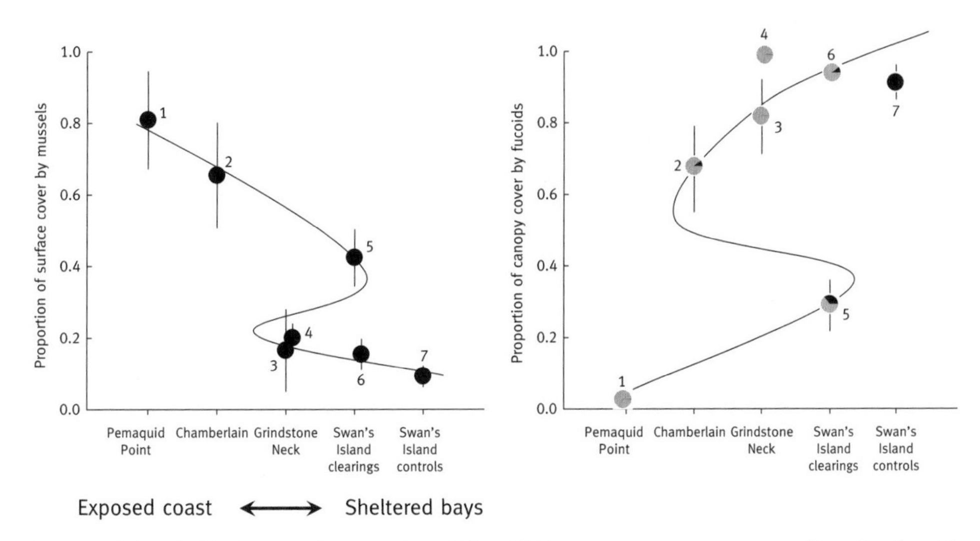

Figure 4.4 Thresholds in cover by mussels and fucoid algae across an exposure gradient. For fucoids, the pie symbols give the relative abundance of *Ascophyllum nodosum* (black) versus *Fucus vesiculosus* (gray). Points without error bars have standard errors smaller than the size of the symbol. Points 1, 2, and 3 are calculated from Menge (1976, his Figures 2 and 4). Data in his figures are averages over time, and so the averages used here are the grand averages across a set of averages. By definition, the standard deviation of a set of averages is the standard error. Points 4, 5, 6, and 7 are from data used in Petraitis et al. (2009) and from samples taken at a single time (i.e. averages and standard errors over space). Point 4 is the average of the benchmark data from Grindstone Neck, and point 7 is the average for the control plots. Points 5 and 6 are averages from communities in the large experimental clearings (>2 m) that were identified as either mussel beds or fucoid stands; 27% were identified as mussel beds and 73% as fucoid stands. Clearings identified as mussel beds by Petraitis et al. (2009; their Figure 2b) were a mixture of two types—either barnacle beds with a few mussels or mussel beds with some barnacles, and here barnacle beds (clearings with <12% mussels) were dropped. Lines are fitted by eye, and the S-shaped curves are inferred.

The results of Petraitis and his colleagues are consistent with the earlier findings of Menge, albeit with a different interpretation. Petraitis et al. (2009) suggest that the occurrence of mussels and rockweeds as alternative states is based on priority effects. It all depends on which species becomes established first. In large patches, succession becomes uncoupled from the surrounding community, and they note that recruitment and predation depend on clearing size. Barnacles, mussels, and the rockweed *F. vesiculosus* have better recruitment in large, open areas well away from the canopy edge (Dudgeon and Petraitis 2001), and their consumers are common under algal canopies but rare in open areas. Barnacles facilitate the recruitment of both rockweeds (Lubchenco 1983, Kordas and Dudgeon, unpublished data) and mussels (Chipperfield 1953, Menge 1976, Petraitis 1990, Seed 1969). Thus it is race in large open patches as to which species—rockweeds or mussels—will settle and become established first.

Once mussels or rockweeds are large enough and common enough in a new patch, they begin to modify the local environment and act as ecosystem engineers. Established rockweed canopies block the recruitment of mussels and provide habitat and refuge for mussel predators (Bertness et al. 2004b, Lubchenco and Menge 1978). Established mussel beds can resist attacks by predators (Petraitis 1987) and block the recruitment of rockweeds through the production of large amounts of pseudo-feces (P. S. Petraitis and S. R. Dudgeon, unpublished data). These positive feedbacks keep rockweed stands as rockweed stands and mussel beds as mussel beds.

Petraitis et al. (2009) view the link between the rockweed canopy and mussel predators quite differently from Menge and Lubchenco. Menge and Lubchenco see mussel predators as the indirect cause of the presence of rockweeds. Predators eat mussels, which are better than rockweeds at competing for space, and so the elimination of mussels allows rockweeds a foothold. In contrast, Petraitis et al. believe the causal arrow is reversed, and the presence of a rockweed canopy causes an enhancement of predation. Rockweeds provide habitat for predators to arrive and it is their presence in established rockweed stands that keep mussels at bay.

4.4 Concluding remarks

The experiments of Morley, Chase, and Petraitis and colleagues are good examples of how the theory of multiple stable states can be tested. All three examples meet some of the four general conditions: (1) that experiments must be done at "the very same site," (2) the outcome must be "different" communities, (3) the communities must be "self-replicating," and (4) perturbation of the site or the original community should be a pulse disturbance (Peterson 1984, Petraitis and Dudgeon 2004). However, none of them came close to meeting all seven aspects of the template of a good experimental protocol for detecting multiple stable states that we discussed in Chapter 3.

The criterion of having "different" communities is problematic because we first need to define how different is different. Natural variation between and within the community states makes it extremely difficult to set clear distinctions between multiple states. Petraitis et al. (2009) argue that to do so requires an independent assessment of natural variation so that benchmarks can be developed. These benchmarks are then used to classify experimental outcomes.

This was done in Petraitis et al.'s experiment. They collected data from 52 sites in the Gulf of Maine spread over a straight line distance of 186 km. The data were sorted by community type (rockweed stands or mussel beds) and used to estimate a discriminant function. The function was then used to classify the experimental clearings and the control plots.

Neither Morley nor Chase made an independent assessment of community states. For Morley's example it may be possible to decide if the bimodal pattern of live weights represents two distinctly different groups. Data on livestock production are and have been widely collected, and with these data an independent assessment could be done.

However, Morley did not give actual weights—in fact his Figure 1 does not have ticks or units on the axis for average weights of ewes. While there are clear statistical differences among plots and good evidence for a bimodal distribution, we cannot evaluate if the statistical result has biological relevance.

Chase's experiment has similar issues. Chase found significant bimodal patterns in biomasses of plants and snails in the presence of predators and at intermediate levels of enrichment (Figure 4.2). Chase does not, however, include information on the natural variation in biomasses per unit volume. We are unable to assess if the results that show significant statistical differences seen in his experiment are in agreement with patterns seen in nature. In his defense, this may be a moot point since patterns in biomass and species composition seen in experimental tanks often match natural patterns only qualitatively.

Morley, Chase, and Petraitis perturbed their systems in different ways. In Morley's example, a parameter—the stocking density of ewes—is varied, and this can be clearly seen if we look at his example within the context of the grazing model discussed in Chapter 2. Stocking density equals p, and changing the stocking density is equivalent to shifting the value of p in Equation (2.10). An increase in stocking density decreases the amount of vegetation available (i.e. N in Equation 2.10) and lowers the rate of biomass consumed by each ewe (i.e. $Nh(N)$; see Equation 2.11). The rate of consumption depends on the amount of vegetation at equilibrium. As we saw in Figure 2.2, there can be one stable equilibrium point for amount of vegetation at low stocking density and the possibility of two stable equilibrium points at high stocking density (in Figure 2.2, intersection A at low density, and B and D at high density). It is reasonable to assume that the average weight of ewes is a function of the rate of consumption, and thus the plot of Morley's example (Figure 4.1) is equivalent to what we saw in Figure 2.2.

Chase and Petraitis did pulse perturbations of stable variables. Chase varied the starting densities of snails, and Petraitis and his colleagues removed rockweeds and other species. Chase also varied the level of enrichment by adding different amounts of nutrients at the start of the experiment. Controlling the amount of nutrients is a press perturbation and can be viewed as tuning a parameter. The clearing of rockweeds by Petraitis and his colleagues was done to mimic the effects of ice scour. This pulse perturbation was done only once and clearly matched the natural disturbance of ice scour both in size and in timing. The experimental clearings were scraped during winter and were well within the range of clearings that are created naturally by ice scour.

The "very same site" requirement is crucial because different community states could be due to undetected differences in environmental conditions and thus different communities reflect different environments rather than different stable states. However, using the very same site is not without problems. As we discussed in Chapter 3, environmental conditions could change at a specific site, and so the switch to a different state at a single or a small number of replicate sites conceivably could be due to undetected environmental change over time. Thus, compelling tests need to be well-replicated at a large number of sites in the same environment. The assumption here is that replication and randomization of treatments should minimize and randomize the effects of undetected environmental variation.

Overall, all three studies appear to met the requirement of starting with the same "site" through replication. The three experiments that we examined have good levels of replication and, in some cases, randomization of treatments. The total sample sizes were 36 for Morley's example, 48 for Chase's mesocosm experiment, and 60 for Petraitis et al.'s field experiment. The level of replication, however, at each treatment level varied because of the number of treatments involved. Both Morley and Petraitis et al. used 12 replicates per treatment level. Morley provided no information on the placement of the replicate plots and Petraitis et al. spread their replicates over four bays with three sites nested within each bay. It is likely that both studies had good control over undetected environmental variation given the large number of replicates per treatment. Chase used a $3 \times 2 \times 2$ factorial design and so had only four replicates per treatment combination. However, the problem of undetected variation is less likely to be a problem in his study because he used experimental tanks and stocked each tank with a mixture of water and organisms from multiple origins. This should have homogenized the starting conditions.

The requirement of demonstrating stability of the system, which Peterson calls "self-replicating," is difficult if not impossible, and Morley, Chase, and Petraitis tackle the issue in different ways. Morley was concerned with the stability of pasture conditions in the face of environmental variation that in turn led to constant and high levels of livestock production. This is what ecologists currently call resistance or the ability to resist external disturbances (Grimm and Wissel 1997). Morley's demonstration of stability is simply the observation that particular practices for pastures and livestock give consistent and predictable levels of production. Morley, however, cautions that an apparently stable situation can crash unexpectedly and notes what works to maintain stability in one place may not work in another.

Chase used constancy over time as his measure stability (see Chapter 3 for Grimm and Wissel's definition of constancy). Chase repeatedly sampled his experimental tanks over time and when the snail densities did not change from one sampling to the next, he assumed the system had equilibrated and was stable. He realized that he was assuming the system was stable and qualified what he meant by stability by calling the observed constancy "some sort of quasi-stability." He did not test for resistance or resilience which could have been done by slightly perturbing snail densities once the system had reached constancy.

Petraitis and his colleagues used Grimm and Wissel's (1997) persistence as their measure of stability and were careful to use the term persistent rather than stable. Recall that Petraitis et al. (2009) collected data from benchmark sites to make an independent assessment of how rockweed stands and mussel beds differed in species composition and community structure. They sampled sites at which they were they were reasonably certain either rockweed stands or mussel beds had persisted for many years. They then assumed that the community composition and structure in the experimental plots was persistent if it matched what was observed in the benchmark communities. In effect, they assumed that ecological structure predicts function. This is a common approach in ecology and is the basis of many studies on the links between biodiversity and ecosystem functioning. This is a weak approach for determining stability, but it may be the only viable approach when studying communities that are composed of very long-lived species such as rockweeds.

5 Catastrophe theory

We are now in a position to place the well-known conventional views of multiple stable states in the broader perspective of catastrophe theory, which seems to have been overlooked by most ecologists. Catastrophe theory has been widely utilized by mathematicians, chemists, and physicists, but not ecologists, and we will explore why that may have occurred. This viewpoint also allows us to expand our horizons and to examine situations in which one model may include both a single unique equilibrium point and multiple stable states. Ecologists often frame the question of whether a community contains multiple states as an either/or choice between communities with a phase shift with a single smooth threshold and communities with multiple stable states and discontinuous jumps. This view arises from envisioning multiple states in their simplest form and by considering parameters one at a time.

In fact, an ecosystem can exhibit a smooth phase shift under some circumstances and a discontinuous jump under other circumstances. This can easily visualized in a three-dimensional plot of single state variable and two parameters (Figure 5.1). The equilibrium values of the state variable, which is shown on the vertical axis as N^*, depend on the combination of two parameters, represented as a horizontal plane on a and b. The familiar graphs for gradual shifts, threshold shifts, and discontinuous shifts in N^* are nothing more than different vertical slices at different values of b.

There is nothing to prevent us from taking diagonal or even curved slices. The pattern for N^* will depend on the angle of the slice, and there are many possibilities. Clearly all of the various patterns of changes for N^* are not mutually exclusive. Multiple stable states and phase shifts are not an either/or proposition, and multiple stable states could occur at some combinations of a and b but not at others. We should keep in mind that the definition of multiple stable states given in Chapter 2 still applies; the occurrence of multiple stable states is defined by a particular set of parameter values and not by changes in parameter values.

The folds can be visualized by projecting N^* onto a two-dimensional plot of a and b (Figure 5.2). The solid lines are the edges of the folds and their point of intersection is called the cusp. There are three equilibrium points inside the V-shaped notch, and while we usually think of the three being two stable points and one unstable, the reverse of two unstable and one stable is also possible. Outside of the notch there is only one equilibrium point. Discontinuous jumps occur at the cusp and the folds. At these

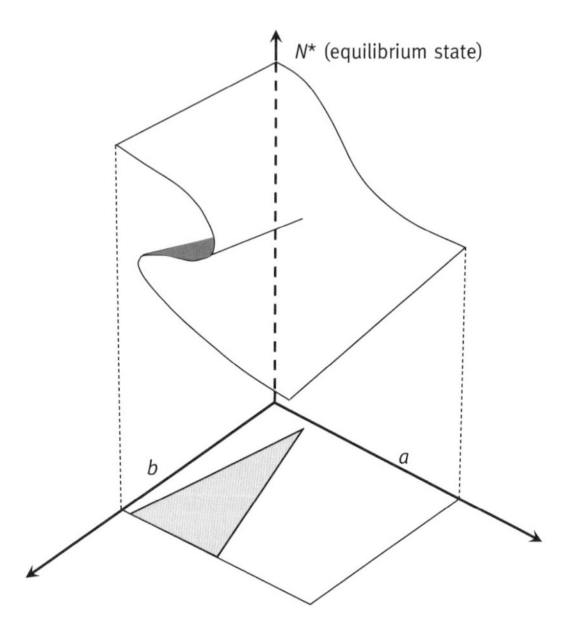

Figure 5.1 Cusp catastrophe showing the shift from a smooth phase shift to the typical S-shaped curve for multiple states. Parameters a and b are as given in Table 5.1. Within the shaded notch there are two stable equilibrium points and one unstable point. The point of the V-notch is a cusp and the lines of indicate the folds. Outside the notch there is one stable point but its position changes as parameters change.

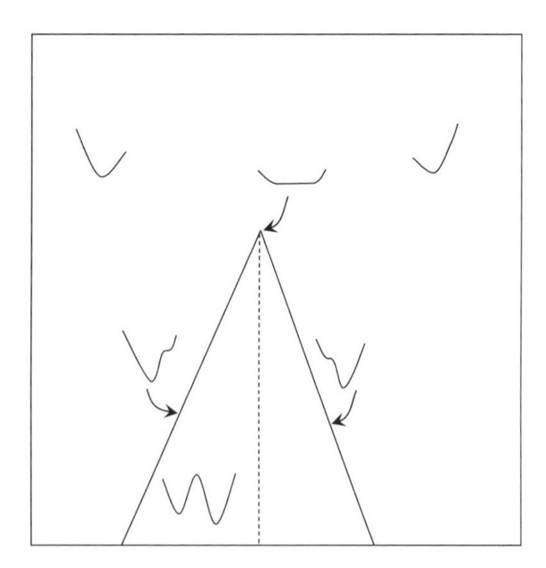

Figure 5.2 Shape of the ball and cup surface on the parameter surface for a cusp catastrophe. On the folds there is a very small region that is completely flat on the steep slope; it is the point at which the discontinuous jump occurs. At the cusp, the surface is flat. The dotted line shows where the two cups are the same depth.

critical points, the shape of the basin is very small and completely flat, which causes the system to jump to the alternative point. Thus the lines denoting the folds and their intersection at the cusp have been called critical points or critical thresholds. Lastly there is one other feature that will be come important later, shown as a dotted line bisecting the notch. Along this line, the two basins have exactly the same depth.

Folds, cusps, and critical points are old ideas in mathematics. Arnol'd (1992) discusses how the American mathematician Hassler Whitney introduced the idea of critical thresholds and described cusps and folds. Arnol'd notes that the terms singularities, bifurcations, and catastrophes have been used interchangeably for this phenomenon. In the appendix to the third edition of his book, Arnol'd discusses what he calls the precursors of catastrophe theory, which go back to 1654. Gilmore (1981) in the prologue of his book also provides a brief history.

5.1 René Thom and an introduction to catastrophe theory

The more infamous proponent of catastrophe theory is René Thom, who developed a topology of catastrophes. Thom's work was first published in 1972 in French and was translated into English by Fowler (Thom 1975). It was popularized for English-speaking audiences by Zeeman (1976) and Saunders (1980). Zeeman uses a different terminology for variables and parameters—using behavior dimensions for state variables and control dimensions for parameters. He discusses hysteresis and uses a figure similar to Figure 5.1 to illustrate how discontinuous shifts can occur in a number of systems. Zeeman also presents a diagram of a ball in a valley and shows how changes in parameters can alter the surface and move the ball to a new position. Zeeman's figure is remarkably similar to Figure 5 in Beisner et al. (2003), which is well-known to ecologists.

Thom's work and Zeeman's popularization fall in between Lewontin's (1969) lament that "[t]he existence of such structural instabilities is very disquieting" and May's (1976) embracement of how simple systems can exhibit discontinuous jumps. Two very accessible books published in English around 1980 (Gilmore 1981, Poston and Stewart 1978) placed catastrophe theory firmly within calculus and did much to clear up the misunderstandings in the fields of engineering and applied mathematics. Ecologists, by and large, have not benefited from these insights. Both books, which include a large number of examples from physics, engineering, and biology, are accessible to anyone with a bit of knowledge of matrix algebra and calculus of several variables.

Ecologists have largely overlooked the relationship between catastrophe theory and modern presentations of multiple stable states. There are a handful of exceptions, which will be discussed in Section 5.2. Ironically, Poston and Stewart (1978) briefly mention Connell's (1961) research on the intertidal zonation of barnacles and how the boundary between *Semibalanus balanoides* and *Chthamalus stellatus* could be an example of a cusp catastrophe. Most ecological models of catastrophes are in either fisheries or rangelands, and all can be traced back to either Jones and Walters' (1976) model for fisheries or Loehle's (1985) model for rangelands. Jones and Walters cite Thom (1975) and Zeeman (1976), and Loehle cites Jones (1977) and Poston and Stewart (1978). Gilmore's (1981) book has been cited over 600 times, but it is not until the late 1980s that the book starts to be cited in the ecological literature (e.g. Deakin 1990, Lockwood and Lockwood 1991, 1993, Ouimet and Legendre 1988). Ouimet and Legendre (1988) provide a statistical methodology for the detection of cusp catastrophes and their work has been cited seven times in the rangeland and shallow lakes literature; Deakin (1990) has been cited twice in the fisheries literature. Lockwood and Lockwood's (1991, 1993) papers have been widely cited within the rangeland literature and are most likely the reason that catastrophe theory has a foothold within that area of ecological research.

The oversight of early descriptions of catastrophe theory by the rest of ecology is probably a product of how modern literature searches are conducted. Thom's and Zeeman's works do not show up if terms such as "alternative states" or "regime shifts" are used in online searches, and it is unlikely most English-speaking ecologists would use "catastrophe theory" in a search for references on multiple stable states. More telling, many of the standard and well-known review articles on multiple stable states since the mid-1970s (Beisner et al. 2003, May 1977, Scheffer et al. 2001, Scheffer and Carpenter 2003, Schröder et al. 2005) do not cite the early descriptions of catastrophe theory (Gilmore 1981, Poston and Stewart 1978, Saunders 1980, Thom 1975, Zeeman 1976). This cuts off the possibility of back-tracking references.

May and Scheffer make a point of dismissing the ideas of catastrophe theory. May (1977) did not cite Thom or Zeeman, but he was clearly aware of the over-blown claims made by some of the early proponents of catastrophe theory and closes his paper with the comment, "Some of these insights can be recast in the language of catastrophe theory, but this is usually *post hoc* window-dressing. It is too often forgotten that catastrophe theory is, strictly speaking, a local theory; we want global descriptions of the dynamics" (May 1977, p. 477). Scheffer (1990), in his first paper on multiple stable states, makes a similar statement.

Thom showed that for systems with one or two state variables and four or fewer parameters, there are seven elementary catastrophes or basic models that contain discontinuous jumps (Table 5.1). The cusp catastrophe is one of Thom's seven "elementary" catastrophes and is the catastrophe that provides us with the familiar S-shaped curve so well known to ecologists. Poston and Stewart (1978) describe how Thom's classification can be extended to include 11 types of catastrophes with five or fewer parameters (also see Arnol'd 1992); Poston and Stewart also provide the computational rules for determining if a system contains a discontinuous shift. A striking feature of all catastrophes is that the function $V(N)$ has multiple valleys and peaks, and so any ecological model based on nonlinear differential equations has the capacity to include catastrophes and multiple stable states.

To make this transparent, let's examine the function V for two well-known ecological models. In ecology, we are interested in rates of change (e.g. dN/dt) and more

Table 5.1 Canonical functions of simple catastrophes involving one or two state variables (x and y). The first seven are Thom's elementary catastrophes, which were named by Thom (1975). Purse and pyramid are names used by Arnol'd (1992). The column labeled ADE gives Arnol'd's A–D–E classification of catastrophes, which is related to simple Lie groups. The terms of each catastrophe function can be subdivided based on their effects, and are known as germ and perturbation. Perturbation is a function of the scaled parameters; k is the number of parameters.

			Function $V(x,y)$					
Thom's name	**ADE**	k	**Germ**	**Perturbation**				
Fold	A_2	1	x^3	$+ax$				
Cusp	$A_{\pm 3}$	2	$\pm x^4$	$+ax$	$+bx^2$			
Swallowtail	A_4	3	x^5	$+ax$	$+bx^2$	$+cx^3$		
Butterfly	$A_{\pm 5}$	4	$\pm x^6$	$+ax$	$+bx^2$	$+cx^3$	$+dx^4$	
	A_6	5	x^7	$+ax$	$+bx^2$	$+cx^3$	$+dx^4$	$+ex^5$
Elliptic umbilic (pyramid)	D_{-4}	3	$x^2y - y^3$	$+ax$	$+by$	$+cy^2$		
Hyperbolic umbilic (purse)	D_{+4}	3	$x^2y + y^3$	$+ax$	$+by$	$+cy^2$		
Parabolic umbilic	D_5	4	$x^2y + y^4$	$+ax$	$+by$	$+cx^2$	$+dy^2$	
	$D_{\pm 6}$	5	$x^2y \pm y^5$	$+ax$	$+by$	$+cx^2$	$+dy^2$	$+ey^3$
	$E_{\pm 6}$	5	$x^3 \pm y^4$	$+ax$	$+by$	$+cxy$	$+dy^2$	$+exy^2$

specifically, the equilibrium solution (e.g. where $dN/dt = 0$). In calculus, these sorts of equations are known as gradient dynamical systems (Gilmore 1981). For the logistic growth model, the relationship between the rate of change for species 1 (i.e. dN/dt) and the function V is

$$\frac{dN}{dt} = -\frac{\partial}{\partial N} V(N; r, K) \tag{5.1}$$

where N is the state variable and r and K are the parameters. Thus from Equation (5.1), we have

$$\frac{dV(N)}{dN} = -\frac{dN}{dt} = -rN + \frac{r}{K}N^2 \tag{5.2}$$

and by integrating we can get function $V(N)$, which is

$$V(N) = \frac{r}{3K}N^3 - \frac{r}{2}N^2. \tag{5.3}$$

The plot of $V(N)$ versus N has a single valley with the lowest point being the equilibrium point, N^*, which is $N^* = K$ (Figure 5.3). This is also where the first derivative equals zero (i.e. $dV(N)/dN = 0$). This is the explicit definition of the surface and is clearly what is implied when ecologists drawn a cup and ball diagram.

Stability depends on the shape of the surface (Box 5.1). If the surface forms a cup, then the equilibrium point is stable; if it is a peak, then it is unstable. We can determine the

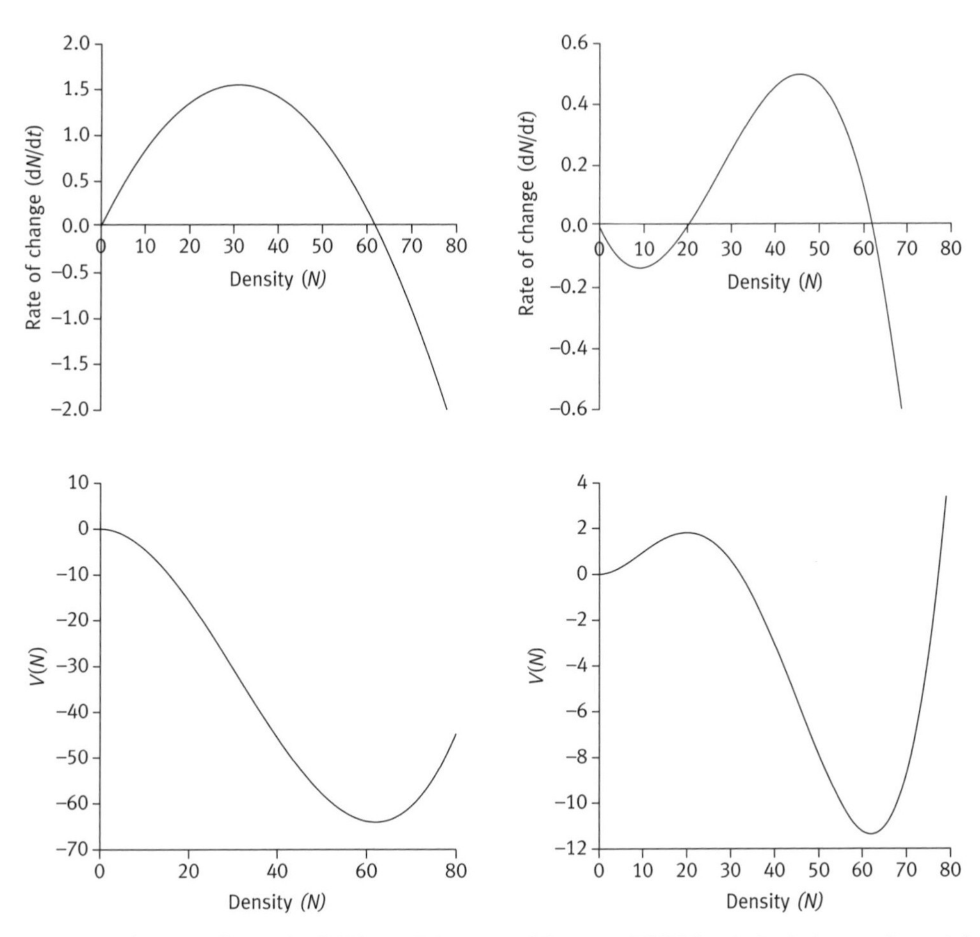

Figure 5.3 The rate of growth (dN/dt) and the potential curves ($V(N)$) for the logistic growth model and the Lewis and Kareiva growth model with an Allee effect. Note that the density at the lowest point on the $V(N)$ curves matches the stable equilibrium point. Logistic growth model: $K = 62$ and $r = 0.1$. Lewis and Kareiva model: $r = 0.1$, $K = 62$. $A = 20$.

shape by taking the second derivative of $V(N)$ and evaluating it at the equilibrium point, N^*. The second derivative tells us whether $V(N)$ is convex and forms a cup or concave and forms a peak. If $d^2V(N)/dN^2 > 0$ at the equilibrium point, then there is a cup and the point is stable. If $d^2V(N)/dN^2 < 0$ at the equilibrium point, then there is a peak and the point is unstable. For the logistic growth model, $d^2V(N)/dN^2 = r$ at $N = K$, and so the equilibrium point is stable.

If the surface of $V(N)$ is very flat near the equilibrium points (i.e. $d^2V(N)/dN^2 \rightarrow 0$), then we have a singularity. This point will become important when we examine some of the characteristics of catastrophes.

Box 5.1 Linear stability analysis

Stability is easily determined by the shape of $V(N)$ at the equilibrium point but requires us to do a couple of things. First, let us center the system at the equilibrium point by defining $n = N - N^*$ and thus

$$-\frac{dV(n)}{dn} = \frac{dn}{dt} = f(n + N^*).$$

Second, using a Taylor expansion, we can rewrite this as

$$\frac{dn}{dt} = f(N^*) + nf'(N^*) + \frac{n^2}{2}f''(N^*) + \frac{n^3}{6}f'''(N^*) + n,$$

where $f'(N^*), f''(N^*)$ and $f'''(N^*)$ are the first, second, and third derivatives of dN/dt with respect to N and evaluated at N^*. For example, $f'(N^*) = d(dN/dt)/dN$ evaluated at N^*. The first term, $f'(N^*)$, is the value of dN/dt at the equilibrium point, which is zero, so this term drops out.

Now thirdly, we make one simplifying assumption. We will assume that shape of $V(N)$ near the equilibrium point (and thus the shape of dn/dt) is smooth and relatively linear. If this is true, then the contribution of $nf'(N^*)$ will be relatively large compared with the contributions of terms with n^2, n^3, n^4, etc. These terms can be dropped with little effect. This is the assumption of linearity. The remaining equation is exponential and thus

$$\frac{dn}{dt} = nf'(N^*).$$

From this equation, it is very easy to see what will happen to the population if we make a small perturbation away from the equilibrium point. If the size of the perturbation at time zero is $n(0)$ then the size of the population at time t will be

$$n(t) = n(0)e^{f'(N^*)t}$$

If $f'(N^*) < 0$, then $n(t) \to 0$ as $t \to \infty$, and thus the population returns to the equilibrium point and the system is stable. If $f'(N^*) > 0$, then the system is unstable.

It is easy to restate these conclusions in terms of $V(N)$. Recalling $dV(N)/dN = -dN/dt$, we get

$$f'(N^*) = -\frac{d^2V(N^*)}{dN^2}$$

and

$$\frac{dn}{dt} = -n\frac{d^2V(N^*)}{dN^2}$$

continues

Box 5.1 (Continued)

where $d^2V(N^*)/dN^2$ is the second derivative of $V(N)$ evaluated at the equilibrium point, N^*. The second derivative tells us if $V(N)$ is convex and forms a cup or concave and forms a peak. If $d^2V(N)/dN^2 > 0$ at the equilibrium point, then there is a cup and the point is stable. If $d^2V(N)/dN^2 < 0$ at the equilibrium point, then there is a peak and the point is unstable. Note that the meaning of inequalities is reversed from what is usually given in ecology textbooks because $dV(N)/dN = -dN/dt$.

A model of dynamics for a single species with an Allee effect will produce two cups separated by a ridge. Lewis and Kareiva (1993) give a model with an Allee effect that is a transparent extension of the logistic model:

$$\frac{dN}{dt} = rN\left(1 - \frac{N}{K}\right)\left(\frac{N}{K} - \frac{A}{K}\right). \tag{5.4}$$

Thus the equation for $V(N)$ is:

$$V(N) = \frac{r}{K}\left[\frac{1}{4K}N^4 - \frac{1}{3}\left(1 + \frac{A}{K}\right)N^3 + \frac{A}{2}N^2\right]. \tag{5.5}$$

The plot of $V(N)$ versus N shows two stable points (Figure 5.3), one when $N = K$ and the other when $N = 0$, and has an unstable point at $N = A$, which is the position of the ridge between the two stable points. The second derivative of $V(N)$ evaluated at the two stable points differs:

$$\frac{d^2V(N^* = 0)}{dN^2} = \frac{rA}{K} \tag{5.6}$$

$$\frac{d^2V(N^* = K)}{dN^2} = r\left(1 - \frac{A}{K}\right). \tag{5.7}$$

The second derivatives are always greater than zero because r, A, and K are positive and A/K is less than one. Thus $V(N)$ is convex and the equilibrium points are stable. In contrast, the second derivative of $V(N)$ evaluated at A gives:

$$\frac{d^2V(N^* = A)}{dN^2} = \frac{rA}{K}\left(\frac{A}{K} - 1\right). \tag{5.8}$$

In this case, the second derivative is always negative, and thus the equilibrium point at the peak is unstable.

For systems with more than one species, the function V is a bit more complex. For example, the Lotka–Volterra model of two-species competition, the relationship between the rate of change for species 1 (i.e. dN_1/dt) and the function V is

$$\frac{dN_1}{dt} = -\frac{\partial}{\partial N_1} V(N_1, N_2, r_1, r_2, K_1, K_2, \alpha, \beta). \tag{5.9}$$

where N_1 and N_2 are the state variables and r_1, r_2, and so on are the parameters. There is also a second equation for species 2. Determining if the system is stable now requires considering both equations and is usually done using matrix algebra. While the procedure for how to do this is covered in many introductory textbooks on ecological theory (e.g. Appendix 3 in Roughgarden 1996), it is worthwhile seeing how this is related to the function V. In general, for a model involving two species we have

$$\begin{aligned} \frac{dn_1}{dt} &= \frac{-\partial V(n_1, n_2, p_1, p_2, ...)}{\partial n_1} = F_1 \\ \frac{dn_2}{dt} &= \frac{-\partial V(n_1, n_2, p_1, p_2, ...)}{\partial n_2} = F_2 \end{aligned} \tag{5.10}$$

where n_1 and n_2 are the densities centered on the equilibrium values (e.g. $n_1 = N_1 - N_1^*$) and with p_1, p_2, and so on as the parameters. Now F_1 and F_2 are already first derivatives of V, and so the second derivatives of V with respect to n_1 and n_2 can be written as the matrix

$$\mathrm{F} = \begin{pmatrix} \frac{\partial F_1}{\partial n_1} & \frac{\partial F_1}{\partial n_2} \\ \frac{\partial F_2}{\partial n_1} & \frac{\partial F_2}{\partial n_2} \end{pmatrix}. \tag{5.11}$$

The key to understanding discontinuous jumps lies with the behavior of the function V close to the equilibrium points. Whether or not an equilibrium point is stable depends on the largest eigenvalue of **F** when evaluated at the equilibrium point. The equilibrium point will be stable if the largest eigenvalue, which we will call λ_1, is greater than zero and unstable if the largest eigenvalue is less than zero. Solving for the eigenvalues requires taking the determinant of the matrix **F** (i.e. det(**F**)), and thus stable points in the conventional sense only occur if det(**F**) $\neq$ 0. Equilibrium points with this property are known as critical points, isolated critical points, nondegenerate critical points, or Morse critical points in mathematics. If det(**F**) = 0, then there is at least one equilibrium point for which one or more of the eigenvalues equals zero. These are the points at which a catastrophe appears—that is places in the system where the state variables "jump" and systems tip or switch from one stable state to another.

This occurs at folds and cusps (Figure 5.2). In cup and model diagrams, the points at which det(**F**) = 0 are seen as a very small and complete flat ledges on which the state variables (i.e. the ball) sits. Points at which det(**F**) = 0 are called non-Morse critical points or degenerate critical points in mathematics. Unfortunately in physics they are known simply as critical points and this has led to some confusion with terminology.

The two basic forms are a fold and a cusp. The fold refers to fold-like pattern created by the two equilibrium values—one stable and the other unstable—of a state variable when plotted against a single controlling parameter. The cusp in a cusp catastrophe refers to the point at which the surface splits into two folds that form an S-shaped curve. The prototypical S-shaped curve seen in dozens of papers about multiple stable states is but a single slice of the cusp catastrophe, which has one state variable and two parameters.

The names become more fanciful and simple geometric interpretations become more tenuous for catastrophes with more than two parameters, but there is an organization to higher-order catastrophes. A butterfly catastrophe contains swallowtails, cusps, and folds, and a swallowtail catastrophe contains cusps and folds. However, the inclusion of more parameters and higher-order terms also means that theory quickly outpaces ecology because the conceptual links between a particular model and observed ecological phenomena are more difficult to specify and justify.

This is not the case with the swallowtail and butterfly catastrophes because they are well within what could investigated by ecologists. The swallowtail catastrophe has three parameters and involves the bending or crimping of a fold (Figure 5.4). The simplest

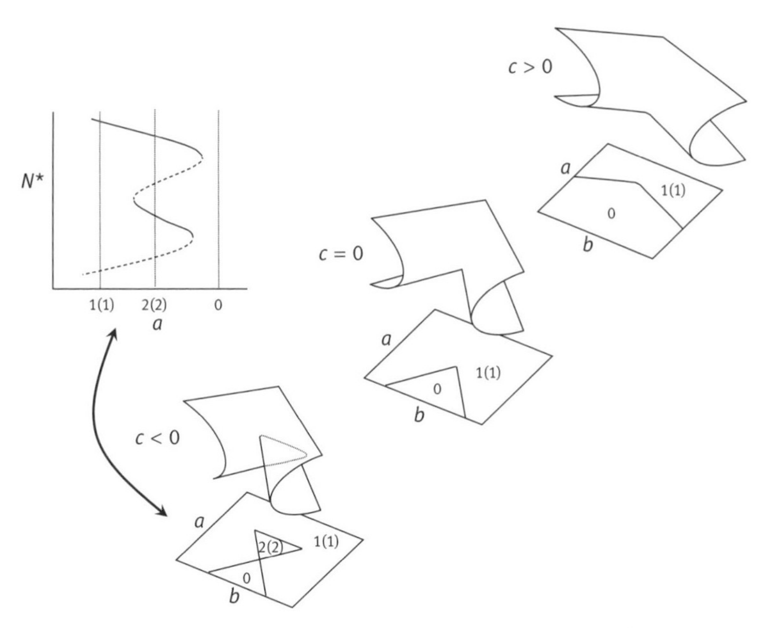

Figure 5.4 Swallowtail catastrophe showing the change in the position and number of equilibrium points as parameters change. Parameters *a*, *b*, and *c* are as given in Table 5.1. The folded surface shows the equilibrium values and the *a*–*b* plane below shows the projection of the folds and cusps. The dotted lines show the fold that is hidden inside; note that the two cusps, which are hidden, are not on the same plane. Numbers on the plane defined by *a* and *b* are the number of equilibrium points; numbers in parentheses are unstable points. The inset shows the plot of equilibrium points versus parameter *a*, and the triple fold that produces two stable and two unstable equilibrium points.

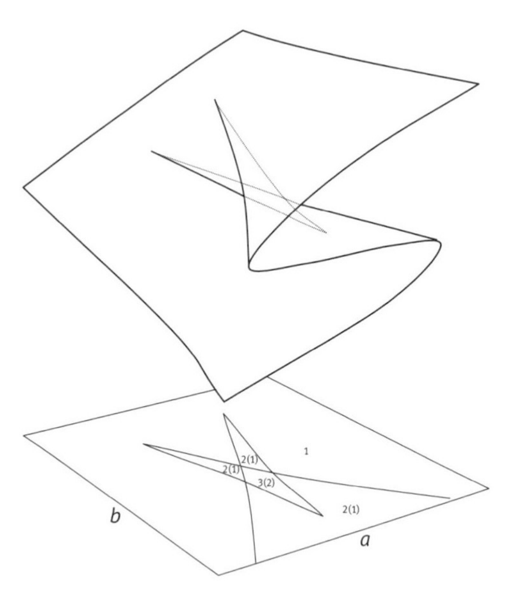

Figure 5.5 The most complex area of the butterfly catastrophe, which has three stable and two unstable points. Parameters *a* and *b* are as given in Table 5.1; parameters *c* and *d* are fixed in this example. The folded surface shows the equilibrium values and the *a*–*b* plane below shows the projection of the folds and cusps. There are three cusps, two of which have one fold visible on the upper part of the surface while the third cusp is completely hidden underneath. Numbers on the plane defined by *a* and *b* are the number of equilibrium points; numbers in parentheses are unstable points.

situation, in which there is one stable and one unstable point, the swallowtail catastrophe looks like a simple fold in a piece of paper. When parameter *c* equals zero, the fold develops a crimp-like point which is a cusp. As *c* continues to decline, the single cusp splits into two cusps that move past each other. The cusps are not on the same plane and thus create an interior region with two stable and two unstable points. The butterfly catastrophe involves four parameters with three cusps and four folds and can present a very complex surface in two dimensions (Figure 5.5). The most interior area has three stable points and two unstable points.

It is worth mentioning Loehle's (1989b) paper, in which he develops a model of grazing that is a butterfly catastrophe (Figure 5.6). Loehle's model has only two "control" variables—grazing and precipitation—but these are included in the model with nonlinear effects on vegetation biomass. As a result, the final model has four parameters and fifth- and third-order terms, as needed for a butterfly catastrophe. Loehle notes that the interior region has three stable equilibria. Loehle's paper is the only ecological example of which we are aware that explicitly discusses a butterfly catastrophe.

Thom's canonical functions are based on one additional insight. Thom knew that a mathematical description of a system in one coordinate system can be smoothly

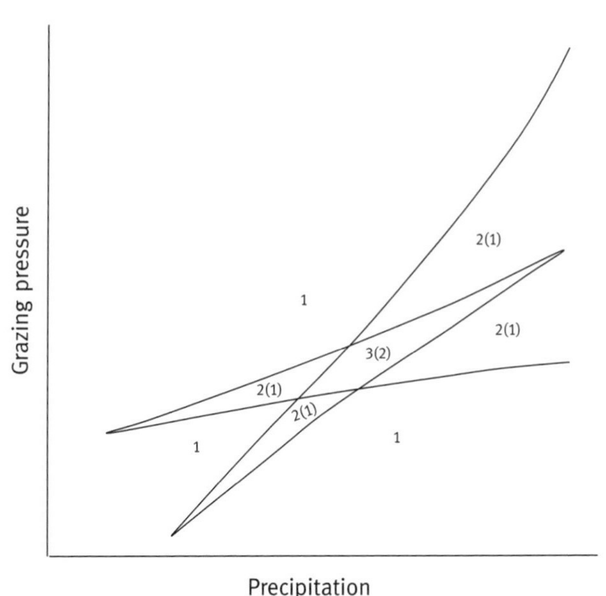

Figure 5.6 Loehle's (1989b) butterfly catastrophe that shows the possibility of three stable states for vegetation biomass under different levels of grazing pressure and precipitation. Numbers on the plane defined by grazing and precipitation are the number of equilibrium points; numbers in parentheses are unstable points. Redrawn from Loehle, C. (1989). Catastrophe-theory in ecology—a critical-review and an example of the butterfly catastrophe. *Ecological Modelling* 49, 125–152. Reproduced with permission from Elsevier.

transformed into a canonical form of a second coordinate system that that splits the transformed variables into two groups—one group that contains variables that involve only Morse critical points and the other group that contains the "bad" variables or those that have non-Morse critical points. The transformation of variables is a standard technique and involves rotating and/or stretching the state space. The change or transformation of variables is called canonical if the shape of the function remains the same. For example, the two-dimensional curve of a saddle can be written as either equation $F(x,y) = y^2 - x^2$ or $F'(x', y') = 2x'y'$ and $F(x,y) = F'(x', y')$ if $x' = x + y/\sqrt{2}$ and $y' = -x + y/\sqrt{2}$. The shape of the curve is identical and only the frame of reference (i.e. the coordinate system) has been changed.

Now suppose a system contains ten state variables $(z_1, z_2, \ldots, z_{10})$ and there are nine "good" equilibrium points and one "bad" critical point in the 10×10 matrix for **F**. What does the system look like at the critical point? Thom showed that

$$F(z_1, z_2, \ldots z_{10}) = x_1^5 + a_1 x_1 + b x_1^2 + c x_1^3 + \sum_{i=2}^{10} \lambda_i x_i^2. \tag{5.12}$$

The summation is the contribution of the nine good variables where the λ_is are the eigenvalues associated with each good variable in the new coordinate system. The four terms involving the one bad variable, x_1, is the catastrophe function $V(x)$. Thus our system has only one troublemaker in the group of ten.

Unfortunately we cannot take advantage of this reduction in number of state variables in many ecological models, which are gradient dynamical systems, are already in a canonical form, and cannot be transformed (Gilmore 1981, Trubatch and Franco 1974). Even so, while standard models in ecology cannot be transformed into the canonical forms, many ecological models of multiple stable states can be viewed as one of Thom's elementary catastrophes.

Some examples of multiple stable states in ecology do not appear to be one of Thom's catastrophes. For example, models of multiple stable states that depend on a Hill function are quite different from Thom's catastrophes. Many of Scheffer's models of multiple stable states, almost all of which use a Hill function, are a good example of this. Scheffer et al. (2003; their Box 3) give a generic model of changes in state, or density N, $\mathrm{d}N/\mathrm{d}t$ as:

$$\begin{aligned}\frac{\mathrm{d}N}{\mathrm{d}t} &= a - bN + f(N)\\ f(N) &= \frac{N^q}{N^q + w^q}\end{aligned} \tag{5.13}$$

where $f(N)$ is the Hill function and q determines the steepness of a threshold (see Equation 2.11). The system will have multiple stable states if the first derivative of $f(N)$ is greater than b, and cannot be written as one of Thom's catastrophes. The same holds for May's grazing model (May 1977), which is quite similar to Scheffer's models.

5.2 Ecological examples

There are very few ecological examples that explicitly use catastrophe theory (Table 5.2). One of the earliest is Jones and Walters' (1976) use of graphical representations and verbal reasoning, similar Zeeman's, to examine how changes in fleet size and fishing efficiency can interact to cause collapses in fishery stocks (see their Figure 3). Jones and Walter consider stock size (the state variable) to be a "fast" variable and efficiency and fleet size (parameters) to be "slow" variables. Jones and Walter cite both Thom and Zeeman. Jones and Walter's model and usage of fast and slow variables was adapted by Collie et al. (2004) and used as part of their explanation of how fish abundances (the fast variable) change in response to internal dynamics and ocean forcing (slow variables). Collie et al. cite Jones and Walter, but not Zeeman. The title of Collie et al.'s paper includes "regime shifts," and so their paper is often cited by others working on multiple stable states. Jones and Walters, who use the older terminology of catastrophe theory, seem to have been forgotten along with Zeeman and Thom.

Table 5.2 Papers with figures of cusp catastrophes as in Figure 5.1. Note that a number of other papers describe cusp catastrophes but do not provide a figure similar to Figure 5.1.

System	Parameters	State variable	Author
Fisheries	Fleet size, efficiency	Stock size	Jones and Walters (1976)
Insect dynamics	Branch surface area, foliage density	Spruce budworm density	Jones (1977)
Phytoplankton	*Anabaena* concentration, phosphate[a]	Algal concentration	Duckstein et al. (1979)
Phytoplankton	Temperature, nutrients	Chlorophyll *a*	Van Nguyen and Wood (1979)
Territorial defense	Distance to nest, reproductive state	Aggression levels	Colgan et al. (1981)
Fisheries and other resources	Slow variables	Fast variables	Rabinovich (1981)
Insect outbreaks	Branch density, predation levels[a]	Larval densities	Casti (1982)
Phytoplankton	External nutrient concentration, uptake rate	Internal nutrient concentration	Kempf et al. (1984)
Tanami Desert, Australia	Spatial variance, temporal variance	Species richness	Saxon and Dudzinski (1984)
Tunisian rangeland	Temperature, grazing by sheep	Shrub density	Rambal (1984)
Freshwater plankton	Not stated	Not stated	Recknagel (1985)
Grasshoppers	Temperature, precipitation	Grasshopper densities	Lockwood and Lockwood (1991)
Rangelands	Not stated	Not stated	Lockwood and Lockwood (1993)
Sahel rangeland	Grazing, rainfall	Vegetation type	Rietkerk et al. (1996)
Wildfire behavior	Windspeed, fuel load, moisture[b]	Fire intensity	Hesseln et al. (1998)
Pinõn–juniper forest	Ground cover, erosion potential	Erosion rate	Davenport et al. (1998)
Fisheries	Fleet size, efficiency	Stock size	Collie et al. (2004)
Semi-arid systems	Structural complexity, degradation	Functional complexity	Turnbull et al. (2008)

[a] Original parameters were transformed into canonical form for the published figure.

[b] The three parameters were combined into two new parameters for the published figure.

Rabinovich (1981) covered much the same ground as Jones and Walters (1976) and introduced catastrophe theory to Spanish-speaking researchers.

Loehle (1985) provides an early example of a grazing model as a cusp catastrophe. His model was explicit and detailed; the rate of plant growth was modeled as a logistic function for plant growth in which the rate of increase varied with rainfall (i.e. the r in the standard logistic equation, as in Equation (5.2), is a linear function of precipitation) and effect of the grazing livestock included a Type II functional response modeled as a Michaelis–Menton curve. Loehle's model contains all the elements found in later models by Scheffer and others. Loehle also makes a very important point, which is sudden jumps are a change in the position of the equilibrium point, not the state variable. Thus approach of the state variable to a new equilibrium point may be very slow even if the jump between points is discontinuous.

Loehle (1989b) provides a comprehensive review of all ecological examples using catastrophe theory between 1979 and 1989. Deakin (1990) has a similar review of catastrophe theory in all areas of biology and includes several ecological examples. Loehle's review is not acknowledged by later researchers (e.g. Beisner et al. 2003, Lockwood and Lockwood 1993, Rietkerk et al. 1996, Scheffer and Carpenter 2003). Loehle also elaborates upon his earlier model of a cusp catastrophe (Loehle 1985) and shows how the fold lines do not need to be a simple V-notch and do not need to be oriented perpendicular to the one of the parameter axes (e.g. see Figure 1 in Rietkerk et al. 1996).

Lockwood and Lockwood (1993) provide a description of catastrophe theory in the context of rangeland management and two competing models of rangeland dynamics. The older view is often called the succession model and assumes that changes from one vegetation state to another proceed in a smooth continuous fashion. The alternative view is the state-and-transition model, which is currently the predominant paradigm in rangelands. The original formulation of the state-and-transition model did not explicitly include jumps (Westoby 1980), but Friedel (1991) suggested that there could also be discontinuous shifts between vegetation states within the state-and-transition framework. The smooth changes can be slow or fast, although it is usually assumed that the fastest continuous change is much slower than a discontinuous jump. Lockwood and Lockwood suggest that both models can be incorporated into a unified paradigm as a cusp catastrophe. Their discussion of catastrophe theory is very limited. They only mention fold and cusp catastrophes although they allude to other types existing.

Rietkerk et al. (1996) cover the same ground as Lockwood and Lockwood (1993) about catastrophe theory in their modeling of the Sahel ecosystem. The Sahel ecosystem lies on the ecotone between the savannas of the Sudan to the south and the Sahara desert to the north. The Sahel contains a mixture of perennial grasses, annual grasses, and annual forbs. There was a dramatic increase in grazing pressure starting in the 1950s, and in periods of high rainfall there appeared to be smooth transitions from perennial grasses to annual grasses and from annual grasses to unpalatable forbs with increasing grazing pressure. The transitions could be reversed with changes in grazing practices. Between 1970 and 1984, there was a long period of drought, and a dramatic and sudden shift from palatable perennial grasses to unpalatable forbs. The shift to forbs was not reversed with the easing of grazing pressure and a series of years with good rainfall (Sinclair and Fryxell 1985, Walker et al. 1981).

Rietkerk et al. (1996) assume there are three distinct communities—perennial grasses, annual grasses, and annual forbs. They assume that rainfall and grazing pressure, which are environmental conditions, can be visualized as the two parameters of a cusp catastrophe. None of the observations from the Sahel are experimental, and so Rietkerk et al. offer, at best, confirmatory evidence that the Sahel dynamics is a cusp catastrophe. They implicitly assume that the system can be described using the single state variable of vegetation response, which implies there is a continuous smooth mapping of the three community types onto this single state variable. Modeling the Sahel ecosystem as a cusp catastrophe also implies that there must be a range of grazing levels at which both perennial grasses and annual forbs occur as alternative stable states and annual grasses are an unstable equilibrium point. There is no evidence that both stable states occur at the same time or that annual grasses can be an unstable state.

As part of their discussion of grassland–shrubland transitions, Turnbull et al. (2008) include a cusp catastrophe diagram and cite Gilmore (1981), Rietkerk et al. (1996), and Lockwood and Lockwood (1993). Turnbull et al. list five of the nine properties of catastrophes, which we will discuss in the next chapter, and link them to specific characteristics of grassland–shrubland transitions.

Finally, and surprisingly, there is some early work on freshwater phytoplankton dynamics which mentions catastrophe theory and has been completely overlooked in the discussions of multiple stable states in lakes over the last 20 years. Modeling of phytoplankton and nutrient dynamics in small lakes using catastrophe theory done in the late 1970s and early 1980s was published in well-known journals (Duckstein et al. 1979, Kempf et al. 1984, Van Nguyen and Wood 1979), yet these early attempts to model multiple stable states receive little or no mention in the better known rediscoveries of discontinuous jumps and hysteresis in lake systems (e.g. Scheffer 1990, 1991, Scheffer and Carpenter 2003, Scheffer et al. 1993, 2001, 2003).

6

Hallmarks of catastrophes

There are nine hallmarks of systems with catastrophes, many of which have been rediscovered by ecologists over the last decade (e.g. Rietkerk et al. 1996). These hallmarks have also been called flags and fingerprints of catastrophes. The hallmarks include not only six characteristics that are well known to ecologists—the existence of two or more equilibrium points for a specific set of parameters (also known as modality), existence of an unstable point in between each pair of stable points (known as inaccessibility), discontinuous jumps, hysteresis, critical slowing down, and anomalous variance—but also an additional three, which are known as divergence, one-jump paths, and nonlinear responses. The hallmarks of critical slowing down, anomalous variance, and nonlinear responses are not unique to systems with catastrophes and multiple stable states. They can also occur in systems with a single stable state, and so cannot be considered as definitive characteristics.

Two hallmarks—discontinuous jumps and hysteresis—do not always occur in systems with multiple stable states. Systems in which the state variables vary randomly due to extrinsic perturbations may not show discontinuous jumps or hysteresis. The occurrence of these two hallmarks depends on how quickly the state variables respond to changes in parameters versus how susceptible they are to random variation. There are two possibilities (Figure 6.1). In the first case, state variables are seen as changing relatively quickly compared with changes in parameters and tend to be immune to the effects of small random perturbations. State variables remain at a particular stable point until changes in parameters allow the state variables to roll to a new equilibrium point. This known as the delay convention, and is the commonsense notion in ecology about how an alternative state is reached if the parameters change (see Figure 1 in Beisner et al. 2003).

The other possibility—known as the Maxwell convention—assumes that state variables are very sensitive to random perturbations and are able to jump to the lowest point on the entire landscape that is defined by the parameters. In ecology, this is usually illustrated by the state variables (e.g. a ball) being pushed over the hill between two valleys, which represent two stable states. This, however, does not fully capture the meaning of Maxwell's convention since the convention assumes that the state variables always and immediately jump to the most stable point. While the Maxwell convention is not usually considered in ecological systems, we can imagine situations in which it may be the norm. Thus before we examine the nine hallmarks of catastrophes, and in

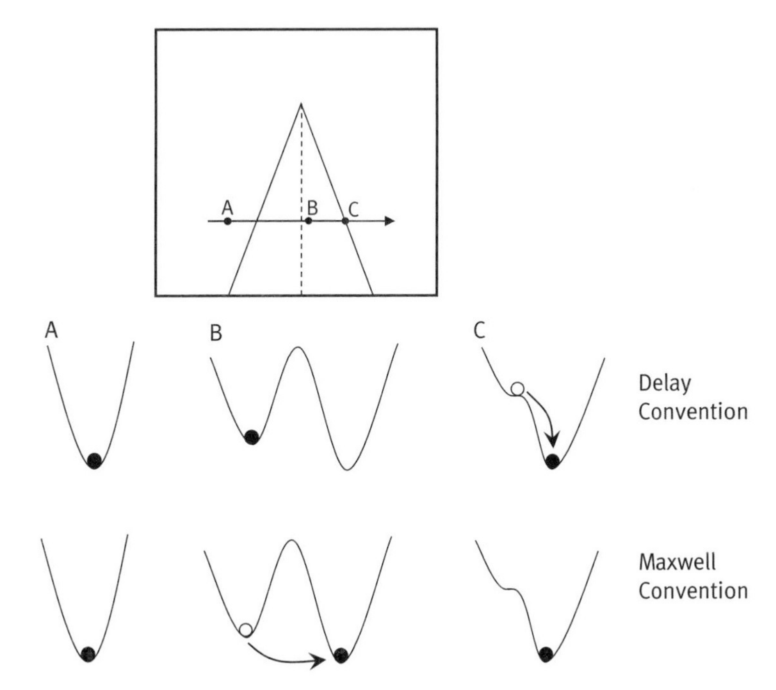

Figure 6.1 Cup and ball diagram of the delay convention and Maxwell's convention. The cup and ball diagrams show the number of the stable and unstable points moving from A to B and then to C. The solid balls show the position of the stable point and the open ball shows when the state jumps to a new stable point. Under the delay convention, there is a jump when the fold at C is crossed. Under Maxwell's convention, the system moves when the dotted line is crossed and the alternative basin becomes lower (i.e. it becomes the globally stable point). Redrawn from Gilmore, R. (1981). *Catastrophe theory for scientists and engineers*. John Wiley and Sons, New York. Reproduced with permission from Robert Gilmore.

particular hysteresis and discontinuous jumps, we must first provide precise definitions of the two conventions.

6.1 Time conventions

The choice between the two conventions depends on how fast parameters change relative to relaxation time and first passage time, which are measures of resiliency. Relaxation time, or return time, which we will denote as T_1, is how long it takes for the state variable to return to a locally stable equilibrium point (Wissel 1984). Relaxation time does not depend on how noisy the system is. In contrast, first passage time, T_2, depends on the probability that the state variables will, by chance, move from their current position and into an adjacent basin of attraction. The chance of moving to an adjacent basin depends on noise in the state variables. Noise, in this case, is random fluctuations in the position of the state variables due to extrinsic perturbations—it is not variation in parameters. Since the noise affects the state variables, we can view noise as a measure of environmental variation that is independent of density. For example, swings in ocean currents that randomly concentrate or deplete phytoplankton densities could be a source of noise. The noisier the system, the shorter the first passage time. First passage time also depends on the shape of the basin of attraction—the deeper the basin, the longer the first passage time. The first passage time is usually much longer than the relaxation time.

Choice between the delay convention and the Maxwell convention depends on how fast a critical parameter changes relative to relaxation time and first passage time. Suppose an important parameter changes over time and let us define that rate of change as dp/dt. The delay convention applies if

$$\frac{1}{T_2} << \frac{dp}{dt} << \frac{1}{T_1} \tag{6.1}$$

and the Maxwell convention applies if

$$\frac{dp}{dt} << \frac{1}{T_2}. \tag{6.2}$$

The rate of change in the parameter p is not the same as the noise that is related to the first passage time. First passage time depends on fluctuations in the state variables unrelated to the changes in the parameters. The rate of change in parameters is a second source of variation and is a measure of how fast a critical parameter, for example birth rate, changes over time.

When might one or the other of these conventions apply in natural ecosystems? For example, imagine the return time for a species with a high rate of reproduction, say a small annual plant, is on the order of ten generations. The first passage time is almost always longer than the relaxation time, so let us assume that first passage time is of the order of 100 generations. This means the delay convention will apply if the rate of change in the parameter p lies between 0.01 and 0.10 generations. We could easily imagine this would be the case if p was affected by day-to-day variation in some environmental factor, for example variation of photosynthetically active radiation for a herbaceous plant. This seems reasonable for many ecological scenarios. However, it is also easy to imagine cases in which the rate of change for the parameter p is relatively slow, and first passage time is relatively short because of large random fluctuations in state variables. For example, changes in population growth rates for trees are likely to change over the order of decades while random variation in densities due to insect outbreaks or fires weather may be over the order of years. In this situation, Maxwell's convention would apply. In order to decide which convention is most appropriate, we will first need precise definitions of relaxation time and first passage time.

Let us first consider the relaxation time, which is the rate of return to an equilibrium point in a system without noise. We know that the rate at which the state of a system returns to an equilibrium point depends on the shape of the basin of attraction

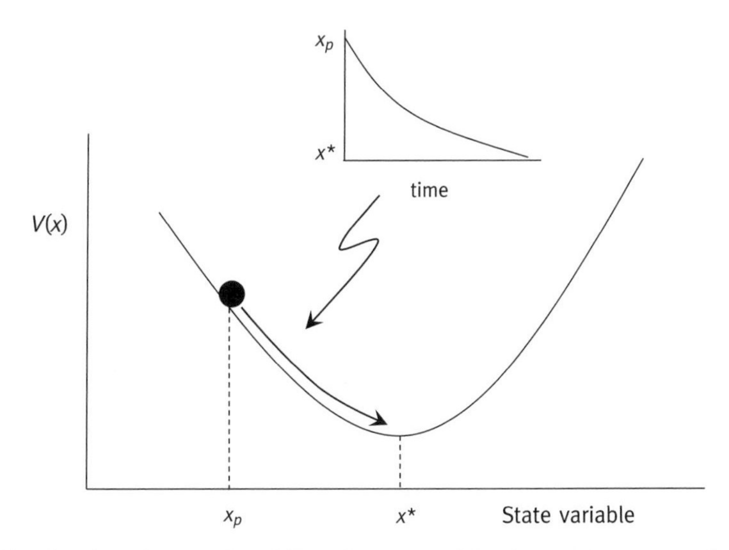

Figure 6.2 Visualization of return time. When the state variable is perturbed to x_p, it "rolls" back to the stable point x^*. The rate of its return is assumed to be an exponential, and is shown in the inset.

(Figure 6.2). Returning to the familiar ball and cup metaphor, the shape of the basin is determined by the parameters. In the simplest case of one state variable, x, the x-axis shows the position of the system. The state variable can be any variable of interest such as the population density of a single species, but to keep the discussion more general we will use x instead of N. The y-axis shows the potential of the system, $V(x)$. The shape of $V(x)$ is defined by the second derivative, $d^2V(x)/dx^2$, which depends on the parameters and the value of x at the equilibrium point. $V(x)$ is at its minimum at the equilibrium point (i.e. $dV(x)/dx = 0$), which is x^*). How fast the state of the system returns to x^* when perturbed a small distance (x_p) depends on the steepness of $V(x)$. The rate of return is measured in terms of the time it takes for the ball to roll back. This measure of time is the relaxation time.

We can provide a precise definition of relaxation time by assuming the second derivative is constant near the equilibrium point (see Box 5.1). Suppose the position of the ball at time t is $x(t)$. If there is a local minimum (i.e. a stable point), then as $t \to \infty$, $x(t) \to x^*$. Now from the assumptions of linear asymptotic stability analysis, we assume that the rate of return is constant if the perturbation is very small. This means that the function for $x(t)$ is an exponential; that is $x(t) = x^*e^{-kt}$ where k is the second derivative of $V(x)$:

$$k = \frac{d^2V(x^*)}{dx^2}. \tag{6.3}$$

For systems with multiple species, k is the largest eigenvalue, which is usually denoted as λ_1.

Relaxation time to a stable point equals $1/k$ and this definition depends on the assumption that movement of $x(t)$ towards the equilibrium point is a random process. Suppose that $x(t)$ is exactly one step away from the equilibrium point and the probability of taking that step is small. Under this model, the probability distribution of the length of time required to make that step is an exponential. This is akin to the rate of decay of radioactive isotopes. Under this model, k is the rate of decay, and it well-known that the average lifespan is $1/k$.

In some situations, the trajectory of one or more the state variables will oscillate during the return to equilibrium and the frequency of the oscillations is approximately the relaxation time (May 1973). This means that the state of the system will cycle through the equilibrium point once every relaxation time interval.

To make this more concrete, let's examine the relaxation for the logistic growth model. Equation (5.3) gives $V(N)$ for the logistic model, and the second derivative of $V(N)$, which determines the relaxation time is

$$\frac{\mathrm{d}^2V(N)}{\mathrm{d}N^2} = -\frac{\mathrm{d}(\mathrm{d}N/\mathrm{d}t)}{\mathrm{d}N} = -r\left(1 - \frac{2N}{K}\right). \tag{6.4}$$

At the equilibrium point, the second derivative of $V(N)$ equals r, and so relaxation time, T_1, equals $1/r$. Keep in mind that relaxation time is a relative measure and is only useful for making comparisons among systems measured in the same time units. The absolute amount of time for return depends on how much a population is shifted away from the stable point. Species with small rates of growth can take a very long time to recover from an unexpected change in population size. This is likely the reason that distinguishing slow recovery from hysteresis is so difficult in long-lived species, which tend to have small intrinsic rates of growth (see comments in Section 3.3 on this problem).

If there is any noise in the system, then there is the possibility that the state variables, which are characterized by the ball, could hop over a ridge and end up rolling into a different basin of attraction. The likelihood of this happening is expressed as the first passage time. For convenience, we will only consider the case of passage time from a local equilibrium point into an adjacent basin, and use the logic developed in Gilmore's (1981) book. More general statements about first passage time from any point are possible but involve more mathematics than is necessary to understand the basic concept.

We will define first passage time as the expected amount of time it takes for the system to move from a local equilibrium point that is not globally stable to the globally stable equilibrium point (Figure 6.3). The locally stable point, $x_{\min}$, which is known as a metastable point, lies in a shallower valley than the globally stable point, x^*. First passage time depends on the amount of noise, the height of the ridge, and the shape of $V(x)$ at the equilibrium points. Let us consider noise first. Noise here is considered to be variation in the state variables and not in the parameters, which determine the shape of $V(x)$. The surface is considered fixed in shape but the position of the ball (i.e. x) on the

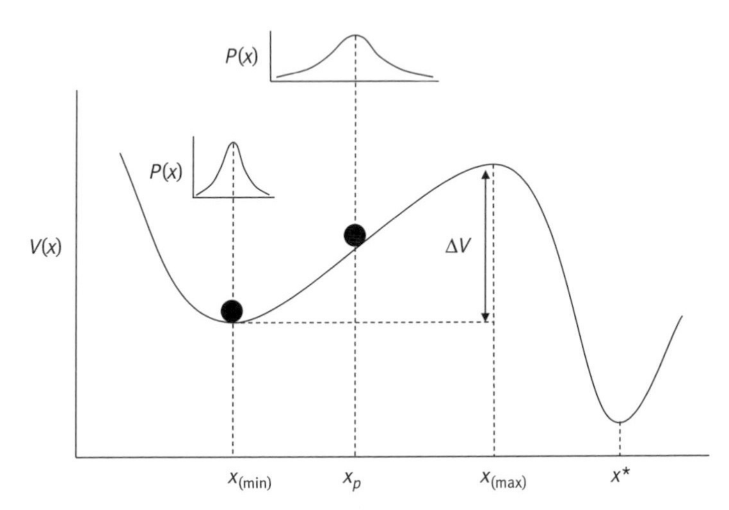

Figure 6.3 Factors affecting first passage time. First passage time is the amount of time it is expected to take for a system to move from a locally stable point, $x_{(min)}$, to the globally stable point, x^*. The insets for $P(x)$ are the probability distributions of x, which are due to the amount of noise in the system. $P(x)$ is always wider away from the stable equilibrium points. Redrawn from Gilmore, R. (1981). *Catastrophe theory for scientists and engineers*, John Wiley and Sons, New York. Reproduced with permission from Robert Gilmore.

surface has some probability distribution. Note that this is quite different from the conceptualization of noise and the switch among multiple stable states presented by others. Many ecological models incorporate noise as random variation in parameters (Beisner et al. 2003, Dakos et al. 2008, Dakos et al. 2010), while we are assuming that noise is extrinsic and causes random perturbations on the state variables. This is an important difference.

A common convention is to assume that the probability distribution of x is due to random noise and thus is Gaussian in shape with a variance related to the diffusion coefficient, D. The larger D is, the more spread there is in the probability distribution. If the state of the system is at a stable equilibrium point (e.g. x^*), then the noise in the system is balanced by the tendency of the system to return to the equilibrium point. If the probability distribution does not change over time then the probability that the state of the system is at a point x near x^* is

$$P(x) = Nexp\left[-V(x - x^*)/D\right] \tag{6.5}$$

where N is the normalization constant. If $D = 0$, then the state sits right on the equilibrium point (i.e. $P(x) = 1$ at x^* and zero for all other values of x).

We can make the meaning of D more understandable by placing it in terms of a one-dimensional random walk. Suppose we start out at a fixed point and we take one step to either the left or right at each time interval. The decision to step to the right or left is determined by a flip of a coin, so we are equally likely to go in either direction. The probability of being m steps away from our starting point after taking n steps approaches a Gaussian distribution when n is large. We now need to express m and n in meaningful units—that is the change in a state variable per unit of time. Let each step have a length δ and let the time interval between steps equal τ. The length of a step is the unit in which we are counting changes in the state variable (e.g. one individual, a hundred individuals, or 100 kg of biomass). Thus the position of the state variable after m steps (e.g. the number of individuals) equals $m\delta$. The time interval τ depends on what time scale we consider relevant for the measurement of a single step (e.g. 1 day, 1 week, or 1 year), and so the total time to take n steps equals $n\tau$. The diffusion coefficient in terms of δ and τ is

$$D = \frac{\delta^2}{2\tau}. \tag{6.6}$$

Thus the diffusion coefficient is a measure of the rate of spread—both in terms of the size of each step and the time interval between steps.

The other two important considerations are the difference in height between the metastable point and the adjacent ridge, and the shape of $V(x)$ at the metastable point and at the top of the ridge. The metastable point is x_{min} and the top of the ridge is the unstable equilibrium point, x_{max} (Figure 6.3). The difference is height is $\Delta V = V(x_{\text{max}}) - V(x_{\text{min}})$. The shape $V(x)$ is determined by the second derivative. Recall that the second derivative tells us how "curvy" the landscape is at a particular point. Also remember that the inverses of the second derivative at $V(x_{\text{min}})$ and $V(x^*)$ equal the relaxation times, T_1, at the two stable points. The flatter the valleys, the longer the relaxation times. The second derivative at $V(x_{\text{max}})$ tells us how flat the ridge is. In systems with two or more state variables (e.g. species), the second derivatives are replaced with the largest eigenvalues at stable and unstable points, which we will denote as λ_{min} and λ^* for the stable points and λ_{max} for the unstable point.

Now we define first passage time, T_2, as the amount of time it takes to move, via diffusion, from the metastable stable point (x_{min}) to the globally stable point (x^*). In equation form

$$T_2 = \frac{2\pi}{|\lambda_{\text{min}}\lambda_{\text{max}}|^{1/2}} e^{\Delta V/D}. \tag{6.7}$$

First passage time declines as the height of the barrier or ridge separating stable points shrinks (i.e. $\Delta V \to 0$), as the amount of noise increases (i.e. D increases), and as the peaks and valleys of the landscape become "flatter" (i.e. λ_{min} and λ_{max} become smaller in absolute value). Surprisingly, first passage time does not depend on the relaxation time to the globally stable point; the depth of the valley containing the globally stable point does not matter.

T_2 can be written in terms of T_1 for the relaxation time to the metastable point and is:

$$T_2 = T_1^{1/2} \frac{2\pi}{|\lambda_{\text{max}}|^{1/2}} e^{\Delta V/D}. \tag{6.8}$$

Note that T_2 will be greater than T_1 when the hilliness of landscape is greater than the noisiness of the state variable (i.e. when $\Delta V/D > 1$).

We are now in a position to make precise statements about when the delay convention and Maxwell's convention apply. As stated at the beginning of the chapter, the delay convention holds if $\frac{1}{T_2} << \frac{dp}{dt} << \frac{1}{T_1}$, and the Maxwell convention holds if $\frac{1}{T_2} >> \frac{dp}{dt}$. In addition, neither applies if $\frac{dp}{dt} > \frac{1}{T_1}$ or $\approx \frac{1}{T_1}$ or $\approx \frac{1}{T_2}$. To make this more concrete, let us examine the relaxation and first passage times for Lewis and Kareiva's (1993) model of single-species growth with an Allee effect (Equation 5.4). Figure 5.3 (bottom right panel) provides the plot of $V(N)$ versus N with stable equilibrium points at K and 0 and an unstable equilibrium point at A. Let $T_1(0)$ be the return time to the metastable point at $N = 0$, and let $T_1(K)$ be the return time to the globally stable point at $N = K$. Thus,

$$T_1(0) = \frac{K}{rA} \tag{6.9}$$

$$T_1(K) = \frac{K}{rA}\left(\frac{A}{K - A}\right). \tag{6.10}$$

The return time to the equilibrium point at K is shorter than the return time to zero, because $K > A$. This is apparent from the difference in the steepness of $V(N)$ at K versus $V(N)$ at 0.

The equilibrium point at K is globally stable, and so the first passage time is defined as the time to move from the metastable point at zero to the globally stable point at K. By definition, $\lambda_{\min} = \frac{1}{T_1(0)} = \frac{rA}{K}$, and $\lambda_{\max}$ is the second derivative of $V(N)$ evaluated at A, which is $\frac{rA}{K}\left(\frac{A}{K} - 1\right)$. Thus, the first passage time is:

$$T_2 = \frac{2\pi T_1(0)}{(1 - a)^{1/2}} e^{\Delta V/D} \tag{6.11}$$

where $a = A/K$. Note that by substituting $a = A/K$ into the equation for the return time for the metastable point, we have $\lambda_{\min} = ra$.

The magnitude of the exponent drives the size of the first passage time and so it is worth taking a closer look at $\Delta V/D$. First recall from Equation (6.6) that D depends on the size of each step (δ) and the time interval between steps (τ). Defining the size of each step as a function of A ($\delta = dA$), we now have

$$\frac{\Delta V}{D} = \frac{ra\tau}{12d^2}(2 - a). \tag{6.12}$$

First passage time depends on the ratio $ra\tau/d^2$. If $ra\tau << d^2$ then first passage time is relatively short, and if $ra\tau >> d^2$ first passage time becomes infinitely long. Thus the choice between the delay convention and Maxwell's convention depends on the rate of increase of the population (r), the relative position of the ridge ($a = A/K$), the time interval between steps (τ) versus the relative size of each step ($d = \delta/A$).

Ecologists usually assume the delay convention to be the norm, but this need not be the case. Let us assume in order for a population with an Allee effect to grow that it must maintain 10% of its carrying capacity (i.e. $a = A/K = 0.1$). Also assume that the rate of increase, r, is 0.1 per year. First passage time will be relatively short and Maxwell convention will apply when $ra\tau << d^2$ or $d^2/\tau >> 0.01$. Recall that d is the relative size of each step ($d = \delta/A$) and τ is the time interval between each step. Even if d is relatively small, say $0.1A$, the Maxwell convention will apply if these small random fluctuations frequent. For example, if the time interval between steps is a week, that is $\tau \approx 0.019$ years, then $d^2/\tau = 0.53$, which is much larger than 0.01.

These are very small rates of change. To put this in terms of r, again suppose d and a are about 0.10, and now assume the carrying capacity of a population is in the millions ($\approx 10^6$). The size of extrinsic perturbations (δ) is then on the order of 10^3 or about 0.1% of the carrying capacity. Moreover, Maxwell's convention will apply if parameters are changing at 1% of the population's intrinsic rate of increase (r). All in all, Maxwell's convention may be the norm for many species even though most ecologists assume the delay convention.

More generally, the choice between the two time conventions depends on the amount of noise and the shape of the landscape (Figure 6.4). The amount of

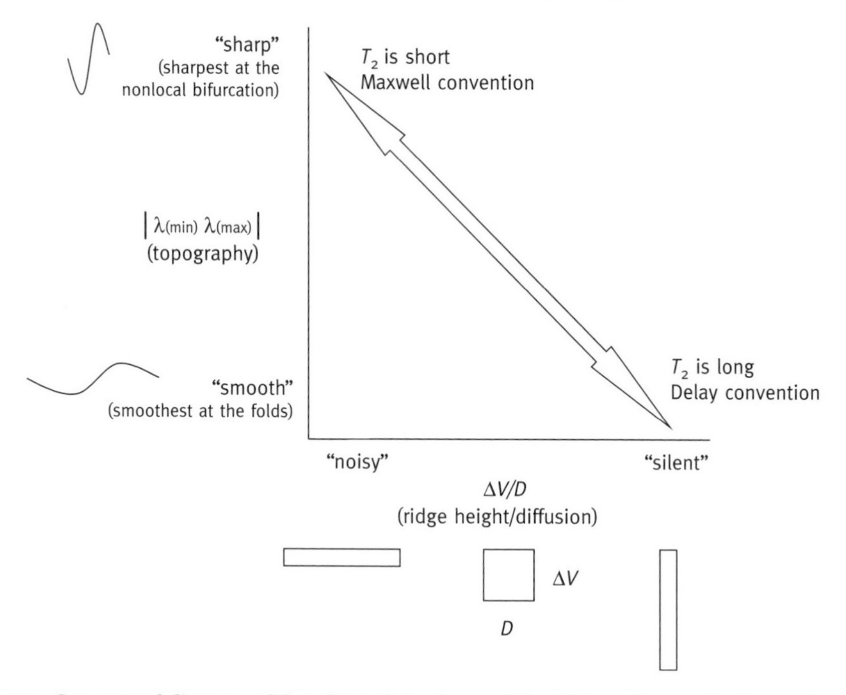

Figure 6.4 Conceptual diagram of the effect of the shape of the $V(x)$ surface and noise on the choice of time conventions. Topography refers to how steep or gradual the contours are. It is possible to have very steep but small hills. The ratio of ridge height to diffusion refers to how big the hills are relative to the noise in the system. Thus a system that is "quiet" and with low hills (i.e. small D and small ΔV) could have the same ratio as a system that is noisy with very tall hills. The choice of conventions depends on the trade-off between topography and relative relief.

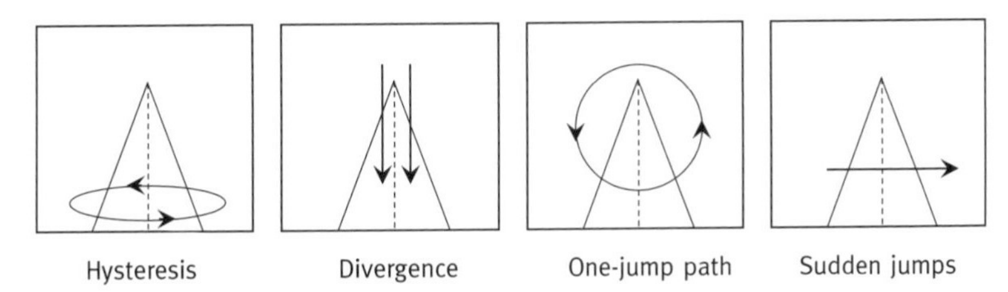

Figure 6.5 Four hallmarks of catastrophes shown on a cusp catastrophe. The arrows show how changing parameters can alter the position and number of equilibrium states. Redrawn from Gilmore, R. (1981). *Catastrophe theory for scientists and engineers.* John Wiley and Sons, New York. Reproduced with permission from Robert Gilmore.

noise relative to the height of the ridges between basins attraction is clearly a factor. The Maxwell convention is more likely for ecosystems in which the landscape lacks much in the way of topographical relief (i.e. ΔV is small) and in which there are strong extrinsic perturbations (i.e. D is large). However, the topography of the landscape on the ridges (λ_{max}) and at the bottom of the basins (λ_{min}) is also important. The "rougher" and "sharper" ridges and bottoms favor the Maxwell convention.

6.2 The flags of catastrophes

We are now in a position to discuss the nine features that can be used to identify catastrophes. These characteristics—which have been called the "catastrophe flags" (Gilmore 1981, p. 157) or phenomena (Poston and Stewart 1978, p. 83 onwards)—were first described in the popular literature by Zeeman (1976) and expanded in Gilmore's (1981) book. Gilmore lists seven; Poston and Stewart list four, including one which is not given by Gilmore. The flags are modality, inaccessibility, sudden discontinuous jumps, divergence, one-jump paths, hysteresis, nonlinear response, critical slowing down, and anomalous variance. These flags have been rediscovered by ecologists over the last decade in one-state-variable models (e.g. Rietkerk et al. 1996), and recently two-species models with critical slowing down have also been developed (Chisholm and Filotas 2009).

The first two flags—modality, inaccessibility—are statements about the condition of the system. Modality is the existence of two or more stable points for the same set of parameters. This flag is thoroughly discussed by Lewontin (1969) and is the implied basis of Peterson's criteria for testing multiple stable states in nature (Peterson 1984, Petraitis and Dudgeon 2004). Inaccessibility refers to the existence of an unstable point in between each pair of stable points and is Lotka's (1956) valley between two mountains and May's (1977) breakpoint curve.

The next four flags—sudden discontinuous jumps, divergence, one-jump paths, and hysteresis—are well known to ecologists (Figure 6.5, see also Lewontin 1969, Lockwood and Lockwood 1993, May 1977, Rietkerk et al. 1996). Sudden jumps are May's discontinuous thresholds in state variables as parameters change past a fold in the typical S-shaped curve, which is found in many ecological papers on multiple stable states. These jumps occur at the folds, where small continuous changes in parameters cause a sudden discontinuous change in the equilibrium values of the state variables. However, this is only true under the delay convention. Under Maxwell's convention, the jump occurs at the nonlocal bifurcation line (i.e. the dashed lines in Figure 6.5) and will be sudden but continuous (Gilmore 1981). The distinction between discontinuous and smooth sudden jumps may not be an issue since in most ecological situations it is likely to be very difficult to distinguish between the two because state variables such as population size do not respond instantly.

Hysteresis is perhaps the best known flag, but it also occurs only in systems in which the delay convention holds (Gilmore 1981). The reason for this can be inferred from the location of the nonlocal bifurcation line. There is only one line, and so the jump under Maxwell's convention occurs at the same parameter value regardless of the direction of change. Our commonsense notions about ecological systems have led ecologists to accept the delay convention as the norm, but there are many ecological systems in which the Maxwell convention may be relevant. In these situations, our inability to detect hysteresis cannot be used to disprove or falsify the existence of multiple stable states.

Divergence refers to the phenomenon that the same change in a parameter can lead to very different states even if there is a very slight difference in the initial parameter value. This means that nearly identical environmental conditions could give rise to very different communities. Note that the flag of divergence is quite different from what is called divergence by many ecologists (e.g. Schröder et al. 2005). The catastrophe flag of divergence is concerned with the divergence of two equilibrium states as a parameter is changed in the same way. In contrast, ecological divergence refers to divergence over time if communities are started in different basins of attraction. The third flag, one-jump paths, is mentioned by Poston and Stewart but not by Gilmore. In systems with folds, it is possible to vary the parameters in such a way that the state of the system makes only one sudden jump before returning to the initial parameter values (Figure 6.5). For ecological systems in which we may have less than perfect knowledge of the actual parameter values or in which we are unaware of how two parameters covary, we may see a system show a sudden jump as environmental conditions change but show a gradual return in the original state as conditions revert.

The last three flags—nonlinear response, critical slowing down, and anomalous variance—are not unique to systems with catastrophes (Gilmore 1981). This is despite the repeated suggestions in the ecological literature that these three flags are reliable indicators of catastrophes and multiple stable states (Dakos et al. 2008, 2010, Rietkerk and van de Koppel 1997, Scheffer et al. 2009). All three will occur if the largest eigenvalue, λ_1, is very small. This includes not only systems with catastrophes where $\lambda_1 = 0$, but also systems with a single stable state in which there is a steep continuous shift in a state variable (i.e. a smooth threshold where λ_1 is very small but not zero). For example,

Scheffer's generic model (Equation 5.12) when $df(N)/dN < b$ will have a threshold and exhibit all three flags but will not contain multiple stable states. Since these flags are not unique to systems with catastrophes, they cannot be considered compelling evidence for multiple stable states.

Since all three flags depend on value of the largest eigenvalue, λ_1, it is worth re-examining stability and what happens when λ_1 approaches zero. Recall that we assumed the shape of $V(x)$ near the equilibrium points to be smooth and the rate of change in x to be constant (see Box 5.1). These assumptions are reasonable when λ_1 is large. Things get more interesting as λ_1 approaches zero. If λ_1 is very small we cannot ignore the "bumps" in the shape of $V(x)$ and the contributions of terms with x^2, x^3, x^4, etc. must be considered. The rate of change near the equilibrium point is then

$$\frac{dx(t)}{dt} = -x\frac{d^2V(x^*)}{dx^2} + \text{higher-order terms} = -x\lambda_1 + \text{higher-order terms}. \quad (6.13)$$

It is these higher-order terms that give rise to a nonlinear response, critical slowing down, and anomalous variance.

Nonlinear response, which is also called divergence of linear response, refers to the behavior of the state variables in response to small changes in parameters. If higher-order terms are not important, then small incremental changes in parameters will cause small linear changes in $dx(t)/dt$. This will not be the case if higher-order terms are important and where small changes in parameters will cause a nonlinear response in $dx(t)/dt$. The divergence from linearity increases dramatically as a cusp or fold is approached. This means the second derivative of $V(x)$ becomes more curved. Rietkerk et al. (1996) incorrectly assert "divergence of linear response... implies that perturbations of the control variables near an edge of the cusp will lead to large oscillations in the state variable" (p. 517). Oscillations may or may not accompany the approach to a cusp or fold, and it requires us to examine specific models to determine which will be the case.

Critical slowing down hinges on the relationship between the largest eigenvalue and relaxation time. Recall that relaxation time is the inverse of the largest eigenvalue. Thus as a system is moved towards a cusp or fold as parameters change, the largest eigenvalue approaches zero and relaxation time will increase dramatically. This occurs because the basin of attraction becomes flatter. This increase in relaxation time slows down the return of the state variables to equilibrium as critical points—cusps and folds—are reached, and this called critical slowing down.

The effects of nonlinear response and critical slowing down are dramatic at the cusps and folds (Figure 6.6). For nonlinear response, the curvature of the second derivative increases, and for critical slowing down relaxation time increases (paths A_2, A_3 and B in Figure 6.6). There will also be a slight increase along path A_1 if there is a smooth threshold shift in the equilibrium value as parameter a changes. This increase can be large in an absolute sense, although it will be small relative to the increases seen at the

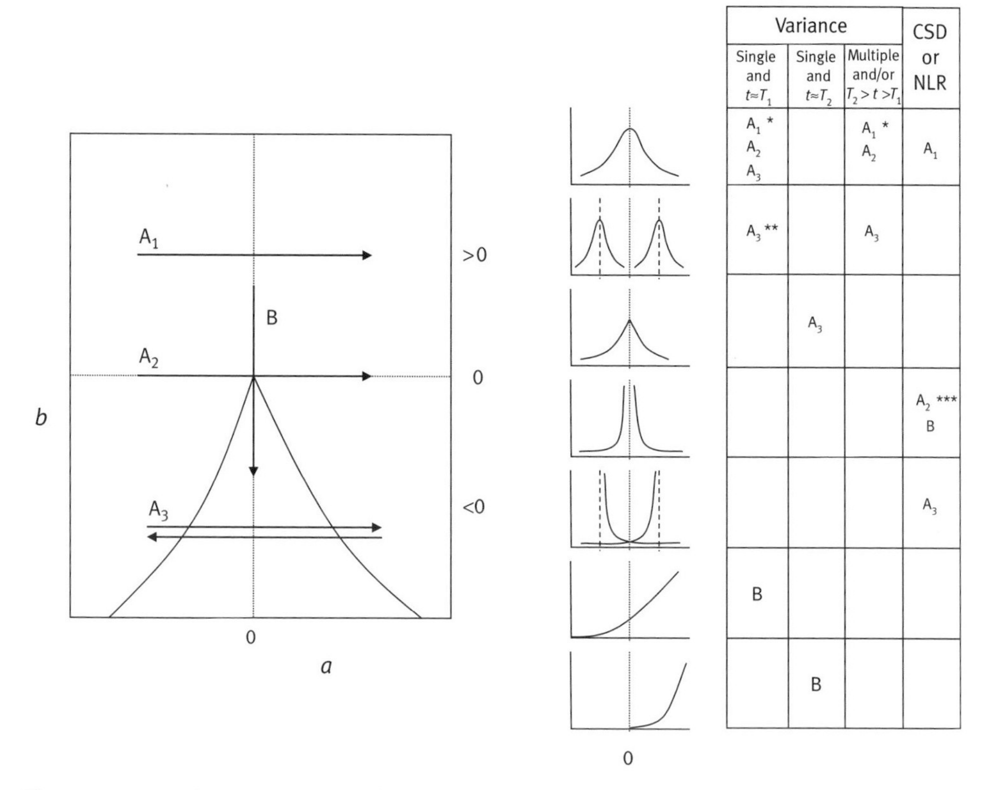

Figure 6.6 Anomalous variance, nonlinear response (NLR), and critical slowing down (CSD) on a cusp catastrophe. There are four possible paths across the parameter space and six possible patterns for anomalous variance, NLR, and CSD. The match between paths and patterns depends on time interval of the experiment or durations of observations (t). On small graphs, the x-axes are either parameter a or b depending path. The y-axes are either $\sigma^2_{T_1}$ or $\sigma^2_{T_2}$ for anomalous variance; relaxation time (T_1) for CSD, or curvature of the second derivative of $V(x)$ for NLR. Some paths are not relevant; for example anomalous variance (e.g. variance is zero) along paths A_1 and B when $t \approx T_2$. Dotted lines on small graphs are either a or b = 0; dashed lines indicate position of folds. *Curve becomes more peaked as b becomes smaller, i.e. $A_3 > A_2 > A_1$. **Curve is bimodal if direction along A_3 is known. ***The y-axis gives the absolute value of the second derivative. Redrawn from Gilmore, R. (1981). *Catastrophe theory for scientists and engineers.* John Wiley and Sons, New York. Reproduced with permission from Robert Gilmore.

folds and cusp. The increase at smooth phase shifts is why the use of changes in critical slowing down to detect discontinuous jumps can give false positives.

The increase in the relaxation time and critical slowing down can cause temporal and spatial autocorrelation. The easiest way to see this is to imagine the position of the ball (i.e. the state variable) at two consecutive time points. If the basin of attraction is very shallow and the ball rolls slowly then the position of the ball at second time point is likely to be very close to its position at the first time point. There will be temporal autocorrelation. Now replace space for time and carry out the same thought experiment, and we will have spatial autocorrelation. While the role of spatial and temporal autocorrelation appears to have been recently discovered by ecologists (Dakos et al. 2010, Donangelo et al. 2010, Scheffer et al. 2009), the phenomenon is well known in other areas in which catastrophe theory has been applied (e.g. see Gilmore's (1981) and Poston and Stewart's (1978) discussions of critical opalescence, and Kubo's (1966) review of the fluctuation–dissipation theorem in physics). In biology, it has long been well known that pattern formation occurs near critical points in reaction–diffusion models (e.g. see Murray's classic work on coat patterns in animals (Murray 1982, 2002, 2003)).

The discussion of anomalous variance is a bit more problematic, because estimates of variances and covariances of ecological processes depend on the spatial and temporal scale over which a set of observations have been taken. Any discussion of anomalous variance or covariance as a catastrophe flag requires us to be very specific about what we are measuring and for how long. The easiest way to examine the effects of spatial and temporal scale is to examine the extremes. First, suppose we have a series of observations either from a single experiment or a single location, and the duration of a single observation is less than the relaxation time, T_1. We will assume that there is no change in parameters, which means the average state remains the same over time. Also assume that the values of the state variables vary from one observation to the next due to small perturbations that are random and external to the system. If the perturbations are modeled as a diffusive process (see Equation 6.5), then the variance over time and the covariance between different state variables, which we will call $\sigma^2_{T_1}$, is proportional to:

$$\frac{D}{\lambda_1} \tag{6.14}$$

where D is the diffusion coefficient. The variances and covariances will very large near a cusp or a fold because this is where the eigenvalue approaches zero (Figure 6.6).

Other extremes involve situations in which the observations are taken either over a very long time interval or from multiple experiments or locations. How long is long enough is relative, and thus is scaled to the first passage time, T_2. Recall this is the expected amount of time it would take for the system to move from a metastable point to the globally stable point via a series of small random perturbations. Now assume we are within a set of parameters for which there are multiple stable states. For a cusp catastrophe, this occurs only within the V-notch. If we observe a single system for long enough, we would expect to see the state variables at all of the equilibrium points, and thus the variance is a measure of the differences among stable points. The same is true for the variance among for a collection of observations taken over a large geographical area. We cannot be sure that the state variables at each location are in the same state, and so the variance will reflect the variance among alternative equilibrium points. In the case of a single-species system in which there are two stable equilibrium points (e.g. x^*_1 and x^*_2), the variance ($\sigma^2_{T_2}$) equals:

$$\frac{(x^*_1 - x^*_2)^2}{2}. \tag{6.15}$$

The two variances differ in magnitude and in their pattern. $\sigma^2_{T_2}$ is always much larger than and tends to swamp out $\sigma^2_{T_1}$. Both increase dramatically as cusps are approached. However, the behavior of the variances differs as folds are approached, and this depends on how the path cuts across the parameter space. First let's examine path A_3, which cuts across the V-notch of a cusp catastrophe (Figure 6.6). Let us also assume we know there is no reversal of the path over the course of our observations. In a single experiment or a short time series from a single location, $\sigma^2_{T_1}$ increases as a fold is approached, and there are two points at which variance is largest; one for each fold. The distribution is bimodal. For a collection of experiments or for a set of observations taken over a very long time period, $\sigma^2_{T_2}$ is largest at the mid-point of the notch. The effect of the single-experiment variance, $\sigma^2_{T_1}$, which is large near the fold, is swamped by the much larger $\sigma^2_{T_2}$, and thus the peaks in $\sigma^2_{T_1}$ at the folds disappear. The patterns for the two variances also differ for path B, which is perpendicular to the A paths and passes through the cusp. For a single experiment, the variance is largest at the cusp and is slightly larger inside the notch than outside. Recall that $\sigma^2_{T_1}$ is inversely proportional to the eigenvalue, thus this variance will be very large at the cusp. In contrast, variance for a collection of observations (or experiments) continues to increase as the cusp is passed and the path enters the notch.

A number of approaches based on temporal and spatial changes in variance or skewness have been developed for detection of thresholds in systems with multiple stable states (Guttal and Jayaprakash 2008, 2009, van Nes and Scheffer 2003). There has also been a suggestion to look for "critical slowing down" in recovery rates following a perturbation of state variables (van Nes and Scheffer 2007). Recovery from a perturbation is relatively slow when the stable state is close to the lip of a bifurcation fold and relatively fast when the state is far for the lip (e.g. see Figure 6.6). Thus van Nes and Scheffer (2007) suggest that a slow down in recovery rates may allow us to predict when a critical threshold is being approached and the pattern of change in recovery rates may provide evidence for multiple stable states. This approach, even though it relies on looking at responses to perturbations, implicitly assumes that we cannot experimentally manipulate either state variables or parameters. Indeed if we could manipulate state variables or parameters then we might as well test for divergence, modality, or hysteresis.

Noise can make detection difficult (Guttal and Jayaprakash 2007), and variation can increase for reasons unrelated to the existence of a threshold (see the discussion of false positives in van Nes and Scheffer 2007). Moreover, changes in variance and/or skewness would be very difficult to track in nature because successful detection requires data from very long time series, which tend to be rare in ecology. Hsieh et al. (2008) suggest that 20 observations in a time series may be too short to be useful.

Second, the confidence limits for estimates of variance, skewness, and correlation are highly dependent on sample size. More observations are always better, but the number of observations should be at least 20 and certainly no fewer than ten. The width of the confidence interval for skewness is much larger than the confidence interval for variance at the same sample size. This suggests that using increases in skewness is likely to be a less useful as an indicator than variance. The outlook for using correlations is complicated by the fact that confidence limits of correlation estimates also depend on the size of the correlation—the smaller the correlation, the larger the interval. Thus if temporal or spatial correlations are relatively weak prior to approaching a fold or cusp, it may be very difficult to detect a change in correlation without extremely large samples.

Third, detection of large variances requires knowing "normal" levels of variability. Thus there must be baseline data that have been collected over a very long time scale or spatial scale and these data must be collected when the system is not near a fold or cusp. This is supported by modeling of changes in variance in systems with multiple stable states (Contamin and Ellison 2009). Contamin and Ellison's simulations suggest that the probability of detecting the approach to a fold is roughly 5% when baseline time series contains 20 observations. To put this in perspective, we would require 20 years of data for an annual species in order to be correct only one in 20 times. It is completely unrealistic to think that rising variance could be used with any degree of acceptable accuracy for long-lived species and ecosystems with long turnover times.

The best study showing changes in variance and autocorrelation used a lab-based system involving *Daphnia,* which has a relatively short generation time (Drake and Griffen 2010). Results were consistent with approaching catastrophe folds, and Drake and Griffen were able to predict approach to the edge of a fold approximately eight generations prior to reaching it. However, this study was extremely well controlled, involved extensive sampling, and had minimal problems with noise due to environmental variability.

Finally remember that anomalous variances and critical slowing down are not unique to systems with folds and cusps. While large variances or autocorrelations may be consistent with the approach to a fold or a cusp, they cannot be regarded as definitive. For example, Dakos et al (2008) showed a change in autocorrelation in climate data and suggested that was evidence for approaching a fold. A closer look at the data showed no change in variance, which should have also occurred as a fold is approached, and strongly suggests the pattern is noise-induced (Ditlevsen and Johnsen 2010). Along the same lines, Hastings and Wysham (2010) show that a large class of plausible ecological models can show regime shifts without changes in leading indicators despite the suggestions that these could be informative (Scheffer et al. 2009).

7

Other modeling approaches

In this chapter we will undertake a brief sampling of some of the models that have been developed by ecologists to examine multiple stable states. The range and diversity of modeling approaches that have been used to investigate multiple stable states is truly astounding, but most approaches tend to fall into one of two broad classes, which we will call conceptual and functional. Conceptual approaches rely on verbal reasoning and visual representations. Visual representations used to describe multiple stable states include cartoon presentations, graph-like diagrams, or box-flow charts. The well-known cup and ball picture of stability leaps to mind as most common cartoon of multiple stable states. The archetypical conceptual model is the state-and-transition model, which was based originally on a verbal argument and a few diagrams (Westoby et al. 1989). Recently the development of state-and-transition models has become very formalized (Bestelmeyer et al. 2003, 2009) and even includes object modeling approaches that were originally developed as engineering software (e.g. UML). Graph-like diagrams can have a state variable plotted against time, an environmental condition, or a second state variable. These conceptual graphs may appear to be similar to graphs derived from mathematical models, but conceptual graphs are often presented without being linked to a specific model. Often conceptual graphs have axes that are poorly defined or lack sufficient information about what is being measured, and units of measurement.

In contrast, functional approaches involve specification of a functional relationship between state variables and parameters; for example $y = f(x)$ or $x(t + 1) = F[x(t)]$. These approaches include not only the more familiar explicit mathematical models common in ecology but also statistical approaches such as path analysis and graphical presentations in which underlying functional relationships are implied. Functional approaches include discrete and continuous models and may contain stochastic processes or spatial components. In some cases, the functional model is not transparent. For example, the rules underlying simulation models are often very explicit, but in some cases modeling is done via a black box using programs such as EcoSim. Here the modeler may be using a program without a complete understanding of the model under development.

Models are often developed as a means to confirm or support the plausibility of multiple stable states in nature. This is usually done by showing a match between a model and a set of observations from the field. The observations can be either a temporal or spatial pattern by state variables, for example densities or biomass of a sentinel species or parameter values such as birth rates, which match the model parameters needed for there to be to multiple stable states.

Using models is the weakest way to demonstrate the plausibility of multiple stable states in natural ecosystems because models are difficult to confirm or disprove. Even the best efforts in modeling often fail to provide clear-cut predictions and alternative scenarios or consider the robustness of particular models against environmental variation in parameters and sampling error (see the comments of Platt (1964) on strong inference). Attempts to confirm models can lead to the classic problem of trying to prove a null hypothesis when there are no clear alternative hypotheses (Burnham and Anderson 2002, Edwards 1976, Underwood 2000). Finally, many modeling efforts have focused on the development of models that show hysteresis, and as we have already discussed multiple states can occur in systems without hysteresis. Even so, there is a strong tradition of looking for a match between models and patterns as a way to show the existence of multiple stable states, and we will return to the use of models to explain patterns in nature in Chapter 9.

7.1 State-and-transition models and other conceptual approaches

Conceptual models are based on the use of graphs, flow charts or diagrams, and verbal descriptions and usually lack a formal structural or functional model. An excellent example is Archer's (1989) conceptual diagram of changes from grasslands to woody shrubs in southern Texas (Figure 7.1). Much of the rangelands of southern Texas was once dominated by perennial grasses but over time these have been converted to wood shrublands that are dominated by mesquite (*Prosopis glandulosa*). The y-axis of Archer's diagram has two attributes—community composition and the time or energy needed to move to from one state to another. Community composition ranges from perennial grasses to woody shrubs. Time to and ease of converting the system from grasses to mesquite is fast and easy, while going backwards to grasses is slow and difficult. The x-axis includes a threshold zone from which the system may go either to grasses or mesquite and a number of environmental conditions that can push the system in either direction. The environmental conditions include grazing pressure, frequency of fire, and probability and rate of establishment of woody plants. It is interesting to note that grazing pressure is not unidirectional and is highest at an intermediate point in the development of a mature mesquite woodland.

Often a conceptual diagram or idea is transformed into a functional model, and the classic case of this is the state-and-transition model (Westoby 1980, Westoby et al. 1989). The concept of states and transitions was first proposed as an alternative to range succession models. Both models were developed as ways to manage the number of grazing animals or stocking levels on rangelands. Range succession models were based on Clements' (1916) and Sampson's (1917) ideas of climax communities, and were first

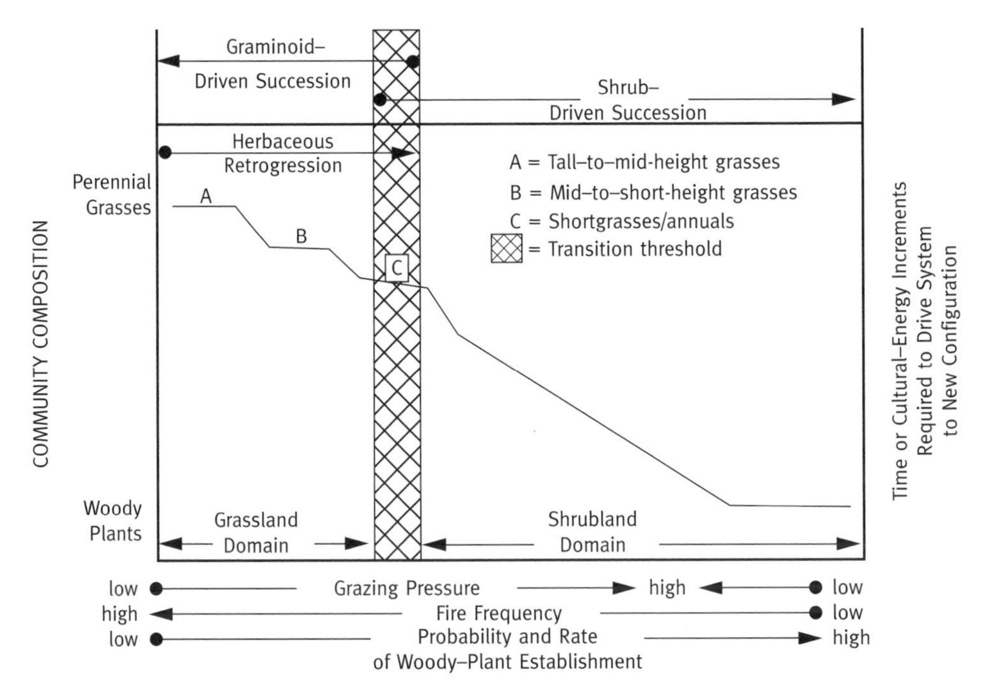

Figure 7.1 Archer's (1989) conceptual diagram for threshold changes and the alternative states of perennial grass and woody plant communities in southern Texas. Archer did not fully define the right-hand axis of "Time or Cultural-Energy Increments," but he wrote in the figure legend, "Once in the shrub-land or woodland domain, the site will not revert to grassland after grazing has ceased, especially if the displaced grasses had originally established under a different climatic regime. Anthropogenic manipulation can alter grass-shrub mixtures, but subsequent succession may result in a rapid return to a community dominated by woody plants." Redrawn from Archer, S. (1989) Have southern Texas savannas been converted to woodlands in recent history. *American Naturalist* 134, 545–561. Reproduced with permission from University of Chicago Press.

formulated by Dyksterhuis (1949). Range succession models assume a single equilibrium point in the absence of grazing and the shift between productive and unproductive land for grazing is smooth, gradual, and reversible without any hysteresis. Productive and unproductive states are viewed as endpoints on a continuum of different stocking levels rather than distinct alternatives. Thus the health of rangelands, that is their state, is a single equilibrium point that shifts as stocking levels are varied. These models formed the basis of rangeland management in the 1940s and 1950s.

Range succession models do not fully explain what is seen in semi-arid rangelands. Grasslands in semi-arid areas of Australia, North America, and Africa can quickly move from being productive to unproductive, and it can be very difficult to reverse the switch. To explain this phenomenon, the state-and-transition approach was developed by Westoby and others (Westoby 1980, Westoby et al. 1989) as a competing idea to the range succession models. The Sahel system, which we discussed in Chapter 5, is

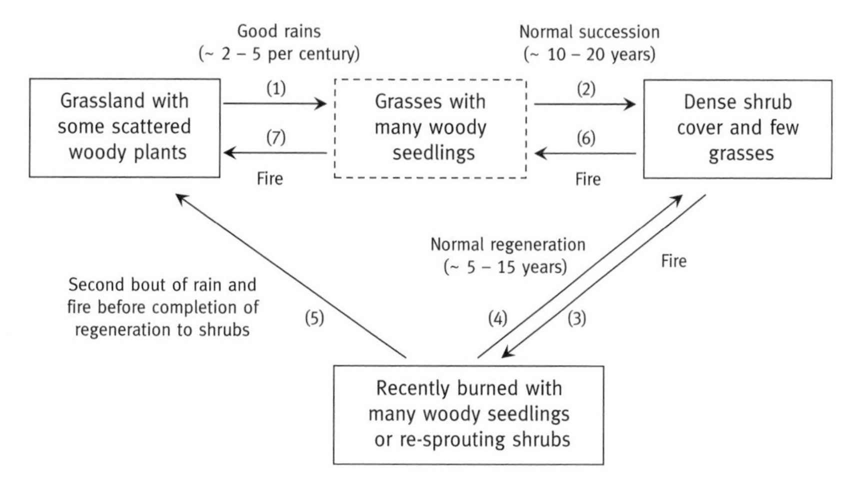

Figure 7.2 State-and-transition diagram for semi-arid grasslands in eastern Australia originally presented by Westoby et al. (1989). Arrows are transitions and boxes are states; the single transitory state is enclosed with a dotted box. Numbers associated with transitions refer to numbering used by Westoby et al. (1989). All fire-driven transitions require prior rains to produce the fuel needed by fires. Grazing and fire suppression by humans slows or blocks the transitions due to fire. Westoby et al. suggested the same model was applicable to semi-arid grasslands in Africa (e.g. the Sahel) and in the south-western USA. Redrawn from Westoby, M., Walker, B., and Noymeir, I. (1989) Opportunistic management for rangelands not at equilibrium. *Journal of Range Management* 42, 266–274. Reproduced with permission from Allen Press Publishing Services.

specific case in which state-and-transition models were used to describe hysteresis in rangelands (Breman et al. 1980, Westoby et al. 1989).

The early state-and-transition models were flow diagrams (Westoby et al. 1989). Boxes represented states of the system and arrows between boxes represented transitions from one state to another (Figure 7.2). Westoby et al. (1989) attempted to explain the transitions between productive grasslands that contained few woody shrubs and dense stands of woody plants that were unusable as pastures for grazing. They suggested the transitions are driven by fire and rainfall, and distinguished between transitions that were rapid or slow. While they used the terminology "alternative states," Westoby et al. were careful to state that they were not developing a model. In fact, they used the phrase "state-and-transition formulation." Westoby et al. went on to suggest that managers could use this formulation as a way to organize the changes seen in rangeland productivity and to offer possible management strategies. They suggested developing three "catalogs." First, list all possible states. Secondly, identify all possible transitions. And finally, consider the opportunities and hazards for management that are provided by the organization of states and transitions. The state-and-transition model of Westoby et al. (1989) holds a dominant position among rangeland studies of multiple stable states, but it is interesting to note that Archer's (1989) conceptualization is strikingly similar and was published 3 months after Westoby et al.'s work.

At about the same time, others were independently developing similar conceptual models with very similar diagrams. For example, Ellis and Swift (1988) proposed that productive and unproductive pasturelands and the switch between the two states was a "persistent nonequilibrium" system. The cycle between good pasture condition, which supports high stocking levels, and poor pasture condition, which can only support low stocking levels, is driven by drought and the slow response of pastoralists to drought. Droughts cause pastures to degrade, but if a drought breaks quickly, the pastures rapidly rebound to good condition with little or no change in the size of herds. However, if a drought persists over many years, pastoralists are slow to respond and stocking levels crash as animals die. Pasture does not recover until the drought breaks and increases in stocking densities lag behind because herds need to time to increase in number. There is slow return to "normal" conditions. Like the model of Westoby et al. (1989), Ellis and Swift's model was completely conceptual. Ellis and Swift also viewed productive and unproductive rangelands as separate states and the transitions between the states could be rapid or slow. Unlike Westoby et al., Ellis and Swift clearly identified these states as persistent but not being held at stable equilibrium points.

State-and-transition models have been applied to other grasslands. Allen-Diaz and Bartolome (1998) examined the transition between sagebrush and grasses in south-eastern Oregon, and Fernandez-Gimenez and Allen-Diaz (1999) undertook a statistical analyses of observational data to examine classical versus "nonequilibrium" models for rangelands in Mongolia. Both studies concluded state-and-transition models, which they considered a nonequilibrium model, provided a better description of these systems. Fernandez-Gimenez and Allen-Diaz (1999) also provide a good overview of the literature and note that managers may be making the same mistakes with state-and-transition models that were made earlier with range condition models.

There have been a number of reviews of state-and-transition models that do not stray from the original conceptualization of a flow diagram with boxes for states and arrows for transitions. For example, Cingolani et al. (2005) provide a tabulation and review of studies in 24 rangeland systems. The analysis of these systems is done on a case-by-case basis and provides a good summary of evidence for the effects of grazing on rangelands. Central to Cingolani et al.'s review is a critique of a conceptual model developed by Milchunas et al. (1988). The model of Milchunas et al. (1988) is presented as a graph of vegetation biomass as a function of grazing intensity. The model suggested that the greatest amount of biomass occurs at intermediate levels of grazing. The graph and verbal reasoning for the shape of the curve have striking similarities to the intermediate disturbance hypothesis (see Petraitis et al. (1989) for a review of the intermediate disturbance hypothesis). Milchunas et al. suggest the shape of the peak or lack thereof depends on resource availability and the length of time that grasslands have coevolved with grazers. Cingolani et al. (2005) add three new aspects for the Milchunas et al. (1988) model that basically relax these assumptions. Like the model of Milchunas et al., the Cingolani et al. model is only presented through reasoning and with conceptual graphs.

In a similar vein, Rodríguez Iglesias and Kothmann (1997) analyzed 29 state-and-transition models in an attempt to determine the number and kind of factors that affect the transitions between states. They grouped the factors into five categories, and not surprisingly grazing was the most common factor and was mentioned in over 30% of the studies. The other four factors were fire, rainfall, "endogenous" characteristics, and human-related activities other than control of fire and woody plants. Endogenous characteristics refer to factors that are intrinsic to the system such as seed banks, soil conditions, competitors, insect outbreaks, presence of invasive exotics. As with Cingolani et al., Rodríguez Iglesias and Kothmann set their review firmly within the state-and-transition formulation.

Other reviews of state-and-transition models also stay well within this conceptual framework. Stringham et al. (2003) review state-and-transition models using only box and flow diagrams and discuss the role that thresholds may have in transitions. Briske et al. (2003) summarize the various paradigms for state-and-transition models with conceptual graphs and suggest that natural systems lie on a continuum of being nonequilibrium versus equilibrium. Unfortunately they equate state-and-transition models as being nonequilibrium, which makes it difficult to see how one could also argue that state-and-transition models are an example of multiple stable states, i.e. systems that contain alternative states that are at equilibrium and are stable. We will return to this misconception in Chapter 8. Briske et al., however, also note that the difference between systems at equilibrium and those that are not depends on spatial and temporal scale. Rodríguez Iglesias and Kothmann (1997) make a far more cogent observation when they comment, "what is a 'state' and what is 'stable' depends on our perception of change."

Conceptual models with boxes and arrows very similar to the state-and-transition formulation also show up in discussions of multiple stable states in coral reef systems (e.g. McManus and Polsenberg 2004, Norström et al. 2009). Both papers review multiple stable states or phase shifts in coral reef systems and provide conceptual models that look identical to a standard state-and-transition diagram. McManus and Polsenberg discuss hysteresis as being involved but give no experimental evidence for it.

There have of course been attempts to translate the conceptual framework of state-and-transition formulation into explicit functional models. May's (1997) presentation of Noy-Meir's grazing model as a single differential equation is an excellent example of this (see Equation 2.2). We have already mentioned the linkage of rangeland dynamics to catastrophe theory (Lockwood and Lockwood 1993, Rietkerk et al. 1996). Rietkerk et al. (1996) suggested that the switch between perennial and annual grasses in the Sahel system, which we discussed in Chapter 5, could be seen as a cusp catastrophe that could be overlaid with a state-and-transition framework (Figure 7.3). Rietkerk et al. (1996) attempt to link catastrophe theory to state-and-transition models for the Sahel ecosystem (see Section 5.2). Perennial grasses, annual grasses, and annual forbs are assumed to be three distinct states and the transitions are due to grazing and rainfall. Slow transitions between the three states occur under high effective rainfall. "Abrupt"

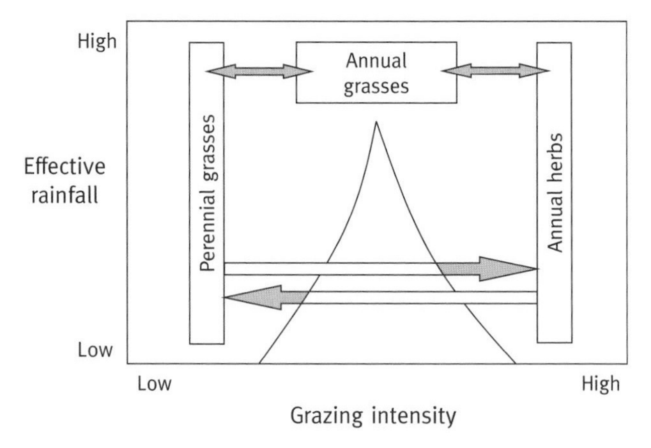

Figure 7.3 State-and-transition model for changes in plant composition for the Sahel ecosystem drawn on top of a cusp catastrophe. Single-headed arrows show discontinuous jumps at the folds and hysteresis. Double-headed arrows show slow gradual changes among plant communities. Redrawn from Rietkerk, M., Ketner, P., Stroosnijder, L., and Prins, H. H. T. (1996) Sahelian rangeland development; a catastrophe? *Journal of Range Management* 49, 512–519. Reproduced with permission from Allen Press Publishing Services.

transitions are sudden discontinuous jumps between perennial grasses and annual herbs; there is no mention of where annual grasses would occur.

7.2 Functional approaches

There are three board classes of functional models for which multiple stable states are possible (Feudel 2008). They are models of weakly dissipative systems, of coupled systems, and of systems with feedback delays. The best known and most widely used functional approaches in ecology to model multiple stable states fall into one of these three categories. Many models can be placed in more than one category, and this is particularly true for models of ecological phenomena. It also does not matter if the model is solved analytically or numerically. This includes models based on differential or difference equations, stochastic differential equations, Markov chains, simulations, and programming packages such as EcoSim. The exceptions are models based on Markov chains. We have already discussed in great detail models based on differential equations that are solved analytically, and so we will focus on other functional approaches as examples of these three classes.

7.2.1 Weakly dissipative systems

Weakly dissipative systems "run down" over time. The diminishing arc of a pendulum is a classic example of a dissipative system. A pendulum will slowly dissipate energy and the arc of each swing will become smaller because of friction. In contrast, a conservative system does not run down as would be the case with a frictionless pendulum. Conservative systems are invariant under time reversal—a video of frictionless pendulum will look the same if run backwards or forwards. We could not tell the difference. Formally, the area or volume of the state variables in phase space is conserved in conservative systems and this result arises from Liouville's theorem. Dissipative systems do not have conservation of volume and irreversibility—a series of arcs of a real pendulum will not look the same if the video is run backwards.

Predator–prey models provide clear examples of both conservative and dissipative systems. The classic Lotka–Volterra model with neutrally stable cycles is a conservative system. The cycling of prey and predator creates an unchanging orbit in phase space and thus shows conservation of area. Adding prey self-limitation causes the predator–prey cycles to dampen. The system runs down to a stable point and thus is dissipative. While weakly dissipative systems have the potential to contain multiple stable states, they do not have to have multiple stable states. For example, the simple Lotka–Volterra model of predator and prey with self-limitation has only one stable point, and as we saw in Chapter 2, in order to turn this model into a model with two stable states requires adding a Type III functional response by the predator.

Weakly dissipative systems with multiple stable states have several features in common (Feudel 2008). The number of stable states scales with the inverse of the rate of dampening. A system in which the approach to the equilibria takes a long time is likely to have more stable states than a system with rapid convergence. Feudel also states, without proof, that in weakly dissipative systems almost all stable states contain periodic orbits and that chaotic attractors are rare.

7.2.2 Weakly dissipative systems with one state variable

All ecological models of multiple stable states that are based on differential equations and involve a single state variable are weakly dissipative systems. This includes May's classic model of grazing and all of Thom's catastrophes involving a single state variable.

As a different example, we will consider a model with a stochastic component. D'Odorico et al. (2006) use stochastic process to switch between tree-dominated and grassland-dominated conditions in savannas. They model tree–grassland dynamics of savannas with a single differential equation for tree biomass. The equation contains two terms: a deterministic term for the encroachment of grassland by trees (i.e. the rate at which grassland biomass is converted to tree biomass) and a stochastic term for the destruction of tree biomass by fire. Fire is modeled as a Poisson process in which the mean is a linear function of tree biomass; fire frequency increases with tree biomass. D'Odorico et al. distinguish between weak, moderate, and strong "fire–vegetation feedback." The steeper the increase in the mean rate of fire frequency with increasing tree biomass, the stronger the fire–vegetation feedback. The system's behavior depends on the feedback. As the feedback goes from weak to moderate to strong, the curve for tree biomass moves from a smooth increase to a smooth threshold-like shift to the S-shaped curve of a cusp catastrophe (e.g. along the axis for parameter b in Figure 5.1).

D'Odorico et al. suggest that arid savanna ecosystems have strong fire–vegetation feedbacks and thus tend to show hysteresis. They only use hysteresis as their criterion for multiple stable states. They go on to suggest that savannas under mesic conditions have weak feedbacks and thus will show smooth gradual or smooth threshold-like shifts because precipitation controls the link between tree biomass and fire frequency.

Recall, however, that hysteresis only occurs if the delay convention applies. For the situation envisioned by D'Odorico et al., the delay convention requires the change in fire frequency (i.e. $\mathrm{d}p/\mathrm{d}t$) to be greater than the inverse of the first passage time from a tree-dominated state to a grassland-dominated status (i.e. $1/T_2$; see Section 6.1). This is likely to be the case under xeric conditions where fire frequency may change quite quickly relative to the inverse of the first passage time. However, under mesic conditions, change in fire frequency may be quite slow, and the Maxwell convention may apply (i.e. $1/T_2 >> \mathrm{d}p/\mathrm{d}t$). Hysteresis would not be observed in these situations even if the system contained multiple stable states.

7.2.3 Examples of weakly dissipative systems with several state variables

Some models of multiple stable states fall into this group, but far fewer than those with a single state variable, for two reasons. First, models of multiple stable states involving two species are often reduced to a model of one state variable in which the effect of one of species on the other is treated as a parameter. Grazing models with vegetation as the state variable and the effect of grazing as a parameter are by far the most common models of multiple stable states. Secondly, ecological models of multiple stable states that involve several species as state variables are most often used by researchers studying coral reefs. In most cases, modeling of coral reefs is done with occupancy models, which are coupled systems and will be discussed in the next section.

Models of weakly dissipative systems with several state variables have been used to model lake systems. For example, Diehl (2007) uses a stoichiometric approach to examine multiple stable states in freshwater producer–grazer systems with a focus on *Daphnia* feeding on phytoplankton. Diehl's model is an extension of the classic MacArthur–Rosenzweig models (Rosenzweig and MacArthur 1963), which address the paradox of energy versus nutrient enrichment in freshwater phytoplankton systems. A stoichiometric approach is often used in these models because primary producers vary in their balance of carbon to nutrients. For example, changes in abiotic resources such as light and nutrients affect not only the growth rate of autotrophs but also their biochemical composition, i.e. the balance of carbon and various nutrients (Droop 1974, Vitousek 1982). Stoichiometric models have been used in other producer–grazer models (e.g. Andersen et al. 2004, Hall 2004, Loladze et al. 2000, 2004, Muller et al. 2001). Diehl's model uses six state variables that are defined by four differential and two algebraic equations involving 27 parameters. It is not a linear system and is weakly dissipative.

Given the richness of the model, both in terms of parameters and state variables, it is not surprising that there are multiple stable states.

Along similar lines, Contamin and Ellison (2009) used stochastic differential equations to examine changes in variance of phosphorus in lakes as the system approaches a fold catastrophe. Recall that one of the hallmarks of systems with multiple stable states is an increase in variance as a fold is approached. Contamin and Ellison were interested in comparing several indictors of variance indictors as a system approached a fold, which they call a "threshold." They based their model on a model developed by Carpenter and Brock (2006) for changes in phosphorus levels in shallow lakes. The Contamin and Ellison model has three coupled stochastic differential equations—one for the concentration of phosphorus in soil surrounding the lake, a second for phosphorus in the lake itself, and a third for phosphorus in the lake sediments. The model includes a Hill function for the recycling of phosphorus from the sediments into the water column. Not surprisingly, the model has a fold catastrophe and as expected indicators of variance increase as the system approaches the edge of the fold.

Fung et al. (2011) developed three very detailed models for coral reefs based on differential equations. In their most complex model, there are three state variables, which are the proportion of the surface occupied by corals, algal turfs, and macroalgae, and 14 parameters. Fung et al. used a genetic algorithm to search the parameter space for areas in which there are multiple stable states (Kelsey and Timmis 2003; immune-inspired algorithm). They found several situations in which there is a cusp catastrophe, and their figures show the typical S-shaped curves of cusp catastrophes. Their results are what we would expect from a complex nonlinear model that is weakly dissipative.

Finally there is model by Dijkstra and Weijer (2005) who used numerical iterative methods to solve for multiple stable states within a well-known nonlinear model of ocean circulation. They rewrote the continuous equations into a discrete form, and their final model is nonlinear and has high dimensionality. Dijkstra and Weijer claimed their approach was the first to provide explicit conditions for bifurcations, although they acknowledged that others have shown the existence of multiple stable states in ocean circulation models. They showed that the results of previous experiments and models can be explained by spatial patterning of freshwater flux, but the model, at best, is only confirmatory of multiple stable states.

7.2.4 Coupled systems

These systems have either linked spatial patches or include more than one state variable. The most obvious example would be the coupled differential or difference equations of competing species or between prey and predator. Coupled systems of this sort encompass Thom's catastrophes involving two state variables and all ecological models of two or more species. Coupled systems also include spatially linked systems and so metapopulation models with multiple stable states (e.g. Shurin et al. 2004) are part of this group. In physics, coupling of identical systems with identical dynamics via diffusion is a common example of a coupled system with multiple states (Feudel 2008) and is very similar to many metapopulation models. The number of multiple states is greatest when the coupling between systems (i.e. patches) is weak and when the coupled systems are not synchronized. For example, multiple stable states are very likely in coupled predator–prey systems in which each patch has slightly different dynamics and the frequency of the oscillations do not match among patches.

7.2.5 Occupancy models as examples of coupled patches

Simulations of occupancy models have been widely used in ecology to study the dynamics of sessile organisms. The approach can be applied to a large range of ecological questions, and within the context of multiple stable states these models have been widely used to examine multiple stable states in coral reefs and, to a lesser extent, in forests and grasslands. Corals and trees are fixed and have reasonably discrete boundaries, and so it is a plausible approach to model these systems with a lattice of spatial cells that are either occupied or not.

The simplest occupancy models contain an array of discrete cells in which each cell can exist in one of two states. A cell could be empty or occupied by a single individual of any species, and the rules for the capture and loss of a cell by an individual vary from model to model. For example, the probability of colonization of an empty cell by an individual of species A is usually assumed to be a function of relative abundance of species A in the system (N_A). In a closed system the colonization probability depends on the proportion of cells occupied by species A, and in an open system the colonization probability takes into account the relative abundance of species A outside of the system. The relative abundance in the pool is assumed to be constant or depend on the per-capita fecundity, depending on the model. The loss of a cell by an individual can occur by death of an individual due to senescence, abiotic effects, or disturbance, which converts an occupied cell to an unoccupied cell, or by the displacement of an individual of one species by an individual of another, e.g. the conversion of a cell occupied by an individual of species A into one occupied by an individual of species B via competition. This displacement can be due to colonization—a colonizer from the pool kills the resident individual—or to over-growth by the individual occupying an adjacent cell. The latter scenario requires knowing the states of adjacent cells and is a feature of spatially explicit models. Finally, more recent models have included environmental variation in site conditions in which the probabilities of colonization, death, and competitive outcome vary spatially and/or temporally.

Models of one or two species with probability rates that are constant or that can described by well-known stochastic processes can often be solved analytically using difference or differential equations (e.g. Horn and MacArthur 1972, Skellam 1951). Models of this sort are also related to meta-population models. Models involving two or more species and constant rates can usually be expressed as Markov models, which can also be solved analytically. Simulations are used to study more complex models, and stability in these models is defined by reaching a steady state. In most cases, researchers do not perturb the system from the steady state to test for resilience or elasticity (sensu Grimm and Wissel 1997).

The connection between occupancy models and more familiar ecological models can be easily seen by recasting an occupancy model into the familiar phase diagram for two-species competition. For example, we can start with Pacala and Tilman's (1994, their equation 6) model and reduce it to a two-species model. The equation for the change in species 1 is

$$\Delta X_1 = -dX_1 + d(m_{11}X_1 + m_{12}X_2) + m_{10}(1 - X_1 - X_2) \tag{7.1}$$

where X_1 is the proportion of cells occupied by species 1, d is the death rate for both species, the probabilities that a cell just vacated by either species 1 or 2 is immediately colonized by species 1 are m_{11} and m_{12}, respectively, $1 - (X_1 + X_2)$ is the proportion of cells unoccupied, and m_{10} is the probability of an unoccupied cell being colonized by species 1. There is an equivalent equation for species 2:

$$\Delta X_2 = -dX_2 + d(m_{21}X_1 + m_{22}X_2) + m_{20}(1 - X_1 - X_2). \tag{7.2}$$

The two equations can be represented as the familiar isocline phase diagram (Figure 7.4). The intercepts are now defined by the probabilities of colonization and the death rate. Just as in the Lotka–Volterra model, the system can have an unstable

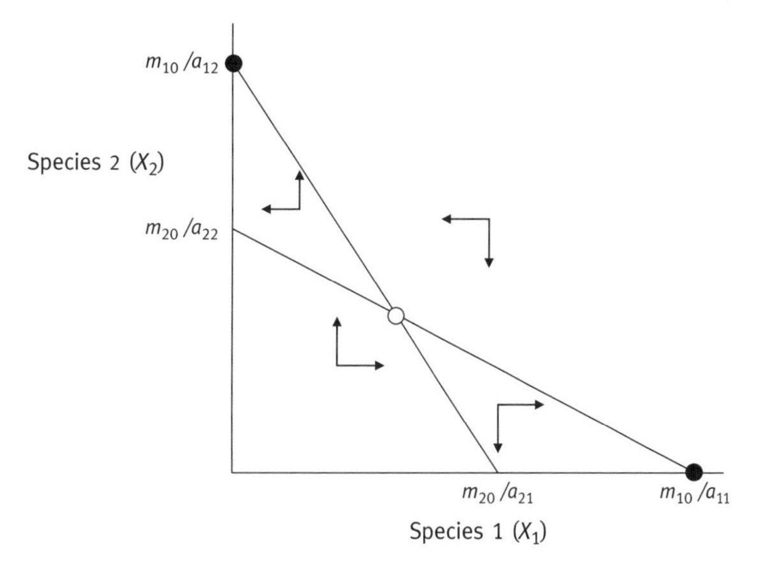

Figure 7.4 Phase diagram for alternative stable states in a two-species occupancy model. The zero isocline for species 1 is $a_{11}X_1 + a_{12}X_2 = m_{10}$ and for species 2 is $a_{21}X_1 + a_{22}X_2 = m_{20}$ where: $a_{11} = m_{10}/(d + m_{10} - dm_{11})$; $a_{12} = m_{10}/(m_{10} - dm_{12})$; $a_{22} = m_{20}/(d + m_{20} - dm_{22})$; $a_{21} = m_{20}/(m_{20} - dm_{21})$. Solid circles are stable points; the open circle is an unstable point.

coexistence in which there are two alternative states. The condition for the unstable coexistence is:

$$\frac{m_{21}}{1-m_{11}} > \frac{m_{20}}{m_{10}} > \frac{1-m_{22}}{m_{12}}. \tag{7.3}$$

Variations of this kind of model have been used to study a variety of questions including dispersal, fugitive species, competition, recruitment limitation, and forest dynamics (e.g. Chesson and Ellner 1989, Chesson and Warner 1981, Horn and MacArthur 1972, Hurtt and Pacala 1995, Ribbens et al. 1994, Sale 1977, Skellam 1951, Slatkin 1974). Hurtt and Pacala (1995) provide a comprehensive listing and a concise summary of the literature starting with Skellam's (1951) paper and through the early 1990s. Most models are used to address the dynamics and equilibria of sessile organisms such as plants or corals because sites are envisioned as spatially fixed, although there is no reason that has to be the case. For example, the total number of sites in these models could just as easily represent the carrying capacity of the system, and the number of occupied sites would then be the current density of individuals. Sites, in this case, are not fixed in space.

Occupancy models have also been used to examine multiple stable states explicitly. Recent modeling of coral reefs is an excellent example of this approach (e.g. Mumby 2009, Mumby et al. 2007). Depending on the model, individual cells are open or occupied by corals, macroalgae, or algal turfs. The spatial structure of the lattices allows organisms to colonize neighboring cells that are open or to overgrow neighboring cells that are occupied. Mumby's simulations contain nonlinear functions to model various processes such as recruitment, spatial expansion of existing colonies, grazing pressure, and competition. Thus it is not surprising that his simulations give rise to multiple stable states. The simulations show corals dominating at low grazing intensity, and macroalgae dominating at high rates with the typical S-shaped curve of a cusp catastrophe at intermediate levels of grazing. The figures in Mumby et al. (2007) and Mumby (2009) show the same pattern seen in May's (1977) graph of his grazing model.

Fukami and Nakajima (2011) examined both stable and transient alternative states and concluded that the occurrence of one did not necessarily predict the occurrence of the other. In some simulations, communities would show very different transient trajectories (i.e. successional pathways) yet converge to a single final state. In others, communities showed minimal differences in transit but then diverged to different alternative states at the end. Fukami and Nakajimi's theoretical results are consistent with Samuels and Drake's (1997) earlier comments that transient dynamics do not predict the final state.

Fukami and Nakajimi assert that strong soil–plant feedbacks contribute to the propensity of alternative stable states, but that need not be the case. They modeled feedbacks through the rules for colonization of an empty cell and the state of adjacent cells. If several individuals "land" on the same empty cell, then the individual from the species that is competitively dominant wins. However, competitive ability is not absolute and is improved if the adjacent cells are already occupied by individuals from the same species. This is the feedback, and because Fukami and Nakajimi were envisioning competition among plants, they viewed the mechanism as being the alteration of soil conditions by adjacent conspecifics that improved the colonization success of new individuals. This is an overly specific view since nearby conspecific individuals could improve colonization success in a number of ways—for example, cooperative behavior among nearby individuals in herds or in nesting colonies. We could go as far as to suggest that Fukami and Nakajimi's feedback is an implicit modeling of an Allee effect since the probability of neighbors in adjacent cells, and thus competitive ability, increases with density.

7.2.6 Systems with delayed feedbacks

Delayed feedback systems are models in which the dynamics of a state variable depends on what happened in the past, for example a system with a time delay in which the dynamics at time t depends on the condition of the state variable at some time in the past. Models of population growth in which the rate of growth depends on population size at an earlier time is a good example, e.g. $dN/dt = f[N(t-T)]$. In addition, models based on difference equations are, by their nature, models with delayed feedback because the state of the system at time t is determined by the state of the system at time $t-1$, i.e. $N(t) = g[N(t-1)]$. May (1973) provides a complete discussion of the structural relationship between differential equations with time delays and difference equations.

Many models of multiple stable states are solved using simulations based on difference equations, and so these models have implicit time delays. Modelers who use simulations to model multiple stable states very rarely inform the reader if the model output is sensitive to the implicit time-step interval that was chosen. Often the choice of the interval appears to be arbitrary. In most cases it is impossible to tell if changing the time-step interval would turn a model that produces multiple stable states into one that has a single stable state. Simulations can also give rise to pathological results; for example, May (1973) notes that limit cycles in linear difference equations are often the result of choosing a "bad" step interval, which imposes a pathological time delay.

Feng et al. (2006) provides a good example of how simulation packages can be used to examine multiple stable states. They used a simulation package to examine alternative states in 36 models based on 26 fisheries. The package is called EwE and is dynamic simulation freeware based on two programs—Ecopath and Ecosim. EwE uses linear difference equations. Feng et al. used EwE to construct mass-balance trophic models for which information on production, diet, efficiency of transfer, and consumption rates was extracted from the literature. In the following discussion, we will assume that changing the time-step interval would have no effect on the results. Each of the 36 models was examined under top-down, bottom-up, and waist-wasp control. Waist-wasp control is a term commonly used in marine ecosystems in which the major consumptive control is by small planktivorous fishes which occupy mid-trophic levels. Each model under the three control scenarios was subjected to a pulse of intense fishing with the target species, which was either a top or mid-level consumer. This gives 36 models and six scenarios. Depending on scenario, Feng et al. reported that 0–23% of the 36 models moved to multiple stages after the release from the pulse disturbance of fishing. They observed both point equilibria and limit cycles.

The outcomes were very specific to particular models; an observation not made by Feng et al. Two scenarios—bottom-up control with either top or mid-level consumers as the target of fishing—never produced multiple states. Eleven models accounted for 23 cases of multiple states with one model in which all four scenarios resulted in multiple states and another seven models in which two of the four scenarios produced multiple states. These results are not surprising given Feudel's (2008) comments that delay-feedback models often have multiple stable states.

7.3 Markov models: functional models with persistent states

Markov chain models use a related approach. In Markov chain models, the state variables are random variables and the transitions are rates of probability. A random variable is a real number that is the outcome of a stochastic process, for example the rolling of seven with two dice. Acácio et al. (2009) present a state-and-transition model as a Markov model with various calculated transition probabilities between states. However, they do not let the model run to a stable solution. The outcome is what they call "persistent alterative states" that are the result of disturbance and arrested succession, and as such are not multiple stable states. Along the same lines, Acevedo (1981) developed a Markov model of tropical forests, and while this model has been cited as an example of multiple states, it is, in fact, a model with a single state.

It is quite easy to have multiple persistent states, or what are called absorbing states, in stochastic and Markov models. The fixation of neutral alleles in the Wright–Fisher model and the Moran model are classic examples. These are genetic drift models in which fixation of one or the other allele occurs via a random walk. Interestingly, this seems to be what Schröder et al. (2005) had in mind in their discussion of random divergence and fixation of alternative community states as a test for multiple stable states.

Stochastic models with multiple absorbing states are, however, not the same as dynamical models with multiple stable states. Arrival at one or another absorbing state is based on a chain of probabilities, and starting at any place in the system, all absorbing states can be reached. It may be highly unlikely, but it is possible. Different absorbing states do not lie in different basin of attraction, and there are no ridgelines to cross—in fact the notion of different basins of attraction has no meaning in simple models of stochastic processes. It is possible to create a stochastic model with several basins of attraction, but in the end this just involves making a stochastic analog to one of the standard dynamical models for multiple stable states (e.g. Contamin and Ellison 2011).

8

Four common misconceptions

There are a number of misconceptions that commonly crop up when ecologists discuss the attributes of multiple stable states in natural ecosystems. We have already touched upon several misconceptions, but it is worthwhile examining four in more detail. These four misconceptions have been stated and restated throughout the literature on multiple stable states, yet they are not necessarily true. They are:

1. Thresholds are always part of systems with multiple stable states.
2. Multiple stable states are central to the state-and-transition concept.
3. In the cup and ball model, the landscape is defined by parameters and the position of the ball by state variables.
4. Characteristics of species and the environment predispose ecosystems to have multiple states.

Other misconceptions pop up from time to time, and three should be mentioned in passing. The first is the notion that systems with multiple stable states are not at equilibrium. This is clearly incorrect since the very phrase "multiple stable states" includes the word "stable," which implies a system at equilibrium. Part of the problem is that conceptualization of equilibrium depends on our biases as observers (Petraitis et al. 1989). Ecological phenomena that occur over large spatial scales and long time intervals are often seen as not being at equilibrium and are thus modeled as transient dynamics. In contrast, small-scale and fast-paced phenomena are often seen as being close to equilibrium. However, the notions of "large" and "long" are often viewed on the human scale. For example the recovery and succession of forests after fires is usually seen as a nonequilibrium process. Yet this viewpoint arises from the fact that trees are long-lived and forest fires are large and infrequent when measured on the human scale. If we step back and consider long time sequences and large areas, then there is nothing to prevent us viewing the effects of fires and the recovery of forests as an equilibrium between the rate of fire occurrence and the rate of forest canopy closure. The rates, however, would be measured in terms of hundreds of years and over thousands of hectares.

Secondly, it is occasionally suggested that multiple stable states must involve a stochastic element (e.g. Bertness et al. 2002, 2004a). This need not be the case. For example, all of the common and well-known models of multiple stable states discussed

Multiple Stable States in Natural Ecosystems. First Edition. Peter Petraitis. © Peter Petraitis 2013.
Published 2013 by Oxford University Press.

in Chapters 2 and 6 are deterministic—not stochastic—models in which stability at equilibrium is the primary concern.

The occurrence of hysteresis in all systems with multiple stable states is the third misconception that deserves a brief mention. This is a very common and pervasive misconception which we have already discussed in detail in Chapter 6. It is clear that systems with multiple stable states in which the Maxwell convention applies will not show hysteresis.

8.1 Systems with multiple stable states have thresholds

Sudden changes are commonly observed in many ecosystems and these abrupt shifts or thresholds are often considered a sign of multiple stable states (Done 1992, Friedel 1991, Knowlton 2004, Laycock 1991, Lenton 2011, McCook 1999, Mumby et al. 2007, Petraitis and Latham 1999, Scheffer et al. 2001). However, there is not a one-to-one correspondence between thresholds and multiple stable states. It is possible to have multiple stable states without thresholds and to have thresholds without multiple stable states. Yet it is a common misconception among many ecologists that thresholds are an intrinsic part of multiple stable states. There have been number of efforts to detect thresholds in time series as a way to infer the existence of multiple stable states. It has even been suggested that in systems with multiple stable states shifts between states are very rapid and are "[t]he key characteristic of a regime shift"(deYoung et al. 2004). In marine systems, this approach has been commonly used in pelagic fisheries or oceanographic studies where the spatial scale precludes the possibility of undertaking manipulative experiments (Andersen et al. 2009, Collie et al. 2004, deYoung et al. 2004, Hare and Mantua 2000, Mantua 2004). When thresholds are detected in time series, or when systems undergo sudden changes, ecologists often turn to developing models of multiple stable states in order to gain some insights into the underlying processes.

This approach has been commonly used in studies of seagrass meadows, coral reefs, and pelagic fisheries. Models of these systems almost always include a threshold. For example, thresholds have been linked to multiple stable states in the switch between seagrass meadows and clear water. Dramatic collapses of seagrass-dominated communities have been observed worldwide, owing to a suite of stressors including eutrophication, coastal loading of sediment and contaminants, rises in sea level and water temperature, increased frequency and intensity of storms, expansion of aquaculture, and spread of nonnative species (Orth et al. 2006). Recovery efforts are often unsuccessful (approximately 30% success rate; Orth et al. 2006), and models of multiple stable states have been developed to explain the apparent hysteresis between seagrass meadows and turbid algal states (van der Heide et al. 2007, Viaroli et al. 2008).

Seagrasses are ecosystem engineers that, in a density-dependent function, reduce water movement and, as a consequence, reduce turbidity (van der Heide et al. 2007). It is assumed that this ecosystem function controls the availability of light, which in turn has a positive feedback in the promotion of plant growth (Scheffer et al. 2001, van der

Heide et al. 2007). The model includes a steep threshold, and not surprisingly the equilibrium densities of seagrass as a function of current velocity show a bifurcation and bistability. The model suggests that once the density of *Zostera marina* is perturbed below a critical density threshold (i.e. crosses the breakpoint curve), turbidity prevents enough stems from recovering to restore ecosystem function.

Viaroli et al. (2008) proposed a similar model to explain the occurrence of seagrass beds (*Ruppia* and *Zostera* spp.) and algal communities (*Cladophora*, *Gracilaria*, and *Ulva* spp.) in Mediterranean lagoons. Here the bifurcation is a function of nutrient loading. The system also has feedback loops, which increases the likelihood of thresholds in parameters. For example, high densities of seagrasses may be able to buffer moderate nitrogen loading through uptake and storage, which is followed by slow decomposition of seagrass detritus (Buchsbaum et al. 1991). This in turn controls algal growth and prevents high densities of macrophytes and phytoplankton that would otherwise reduce light penetration and inhibit seagrass growth. Viaroli et al. (2008) refer to the thresholds encompassing the parameter space where stable states are possible as "thresholds of reversibility," beyond which any perturbation in state variables cannot flip the system to the alternative state.

In much the same way, Mumby et al. (2007) constructed a grazing model of reef coral cover that exhibits a bifurcation as grazing intensity varies. Under strong grazing pressure, as in the Caribbean prior to mass mortality of the urchin *Diadema antillarum*, the system shows a single stable equilibrium of coral dominance. At low grazing pressure, the only stable equilibrium is dominance of macroalgae. At intermediate grazing densities, there are two potential stable states—either corals or macroalgae dominate. Mumby et al. use the term threshold in several ways. They refer to "critical thresholds of grazing and coral cover beyond which resilience is lost," and the threshold is applied to both equilibrium states (i.e. coral cover) and parameter values (i.e. the rate of grazing) at which the bifurcations occur. Mumby (2009) also refers to specific values for the grazing parameter and the unstable equilibria (i.e. the breakpoint curve) as thresholds.

In order to make sense of why this misconception persists, we first need to clarify the differences between processes and states in ecosystems and the usage of the term "threshold" in the context of multiple stable states. Our brief summary does not cover all that has been published on thresholds and multiple stable states, and there have also been numerous well-known efforts to catalog examples of thresholds in multiple stable states (Folke et al. 2004, Knowlton 2004, May 1977, Mumby 2009, Orth et al. 2006, Scheffer et al. 2001). In addition, there have been a number of very comprehensive reviews of thresholds in ecology (Andersen et al. 2009, Briske et al. 2006, 2010, Groffman et al. 2006, Horan et al. 2011, Huggett 2005, Muradian 2001, Osman et al. 2010, Washington-Allen et al. 2010).

In ecology, the term threshold has been applied to sharp transitions in both the ecological processes that control the system and the descriptors of the system itself. Thresholds are perceived as dramatic changes in some aspect of a population, community, or ecosystem, but this very general conceptualization of thresholds involves defining not only what ecologists mean by terms such as "dramatic," but also the context of the question—how ecologists define a population, community, or ecosystem. Ecologists rarely define what a sudden shift is in terms of background conditions, and more often than not "sudden" seems to mean "unexpected" (Doak et al. 2008). We will not fully explore the issue of terminology because there have been a number of recent reviews of thresholds (e.g. Groffman et al. 2006, Osman et al. 2010, Washington-Allen et al. 2010). Here we will focus on the connection or lack thereof between thresholds and multiple stable states.

First we recall the distinction between parameters and state variables, which we discussed in Chapter 2. The parameters of a specific model define the ecological processes that control the system and the state variables are the descriptors of the system itself. Both can vary. Parameters can vary as environmental conditions change—for example, per-capita birth and death rates in an exponential growth model are assumed to change with conditions such as food availability. And of course, the parameters of a model provide the link between environmental conditions and state variables such as population size. As parameters change, so will the behavior, position, and stability of the state variables. Since parameters and state variables can vary, either or both could have thresholds.

However, the relationship between a threshold in a parameter and a threshold in a state variable is not one to one. Let us first consider changes in parameters. As environmental conditions change, parameters may change in a variety of ways. A parameter could show a linear or curvilinear shift or even a steep smooth threshold as environmental conditions change (Figure 8.1; see also Andersen et al. 2009, deYoung et al. 2004, Suding and Hobbs 2009). The exact shape of the curve depends on the underlying physiological and ecological processes that link changes in environmental conditions to changes in parameters. For example, it is easy to imagine that per-capita birth rate would show a threshold response to changes in the availability of food. When food is scarce, we might expect all consumption to be used for maintenance and none for producing offspring. Above a critical minimum of food availability, production of offspring is possible. This sort of scenario could easily give rise to a threshold-like response in per-capita birth rate to increasing food availability. In contrast, it is difficult to conceive how a parameter could have a discontinuous jump and display an S-shaped curve across environmental conditions. This would require some sort of physiological or ecological processes to give rise to two distinct values for a parameter under the same environmental conditions. It is hard to imagine how identical environmental conditions could induce two organisms that were in the same physiological condition to have different per-capita birth rates or different probabilities of death.

We already know from our earlier discussion that there can be dramatic shifts in the equilibria of state variables (see Chapter 5). In systems with a single stable point, state variables at equilibrium could show linear, curvilinear, or smooth threshold-like changes as parameters vary (Figure 8.1b, c). In systems with multiple stable points, we

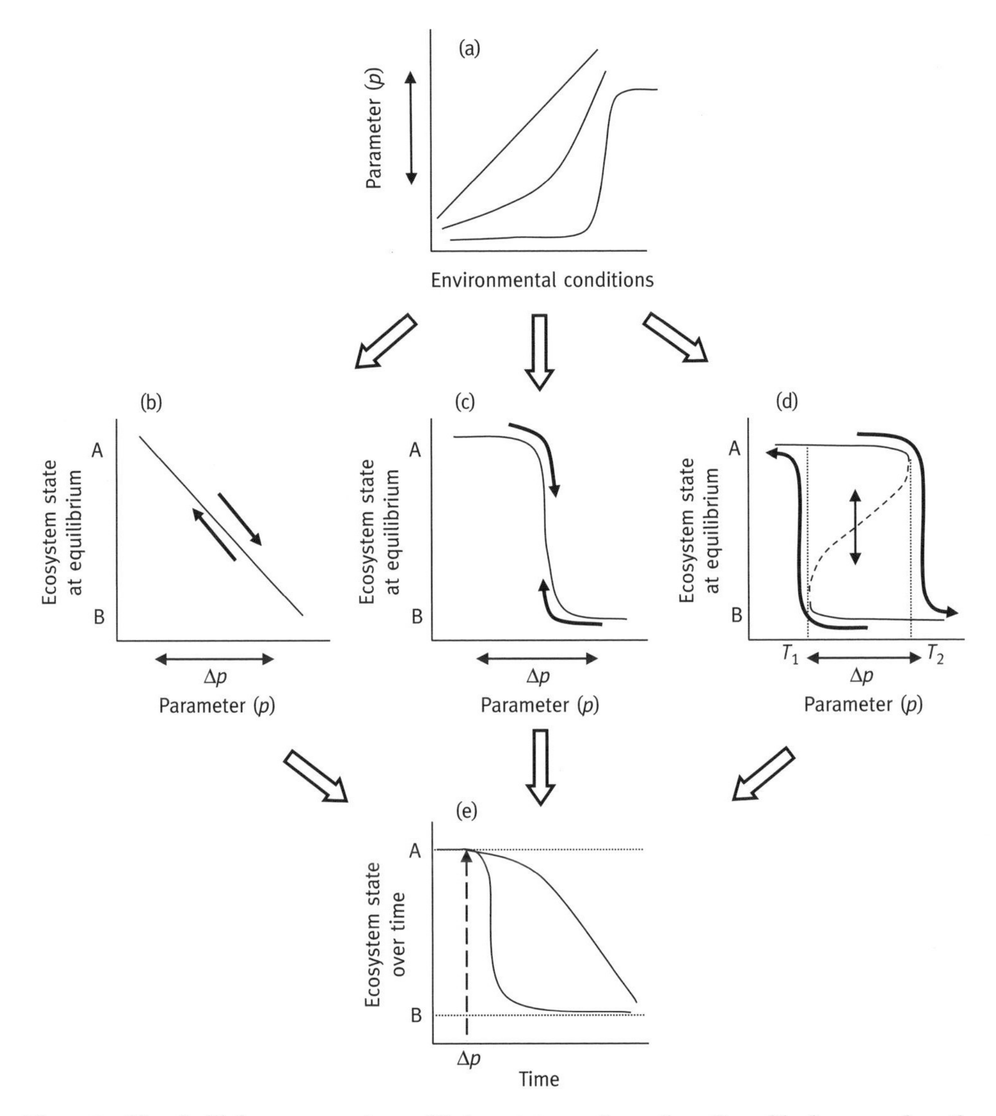

Figure 8.1 Thresholds in processes, in equilibrium states, and over time. Open block arrows show the possible pathways from changes in parameters, in equilibrium states, and over time. Solid arrows show shifts in equilibrium values with changes in parameter values. (a) Parameters can show linear, gradual nonlinear, and steep nonlinear (threshold) responses to changes in environmental conditions. (b) Linear shifts in equilibrium states (e.g. A and B) with changes in parameter values. (c) Steep but continuous shifts in equilibrium states with changes in parameter values. This is a phase shift and not multiple stable states. (d) Steep and discontinuous shifts in equilibrium states from B to A at T_1 and from A to B at T_2 with changes in parameter values. The dashed line is the breakpoint curve and the double-headed arrow a change in state variable without a change in parameter value. (e) Gradual versus rapid changes in ecosystem state over time. The dashed line gives the point at which the system receives a press perturbation in a parameter value. Dotted lines are the equilibrium states (i.e. A and B). Redrawn from Petraitis, P. S. and Hoffman, C. (2010). Multiple stable states and relationship between thresholds in processes and states. *Marine Ecology–Progress Series*, 413, 189–200. Reproduced with permission from Inter-Research Science Center.

will see discontinuous jumps from one stable state to another as parameters vary and move across a fold of a catastrophe.

Thresholds in parameters, however, are not linked to thresholds or multiple stable states in state variables. A threshold in a parameter does not ensure that there will be a corresponding threshold shift in the equilibrium points of a state variable. Conversely, a threshold in a state variable does not imply that here there must be a threshold in a parameter.

So far we have ignored the pace of change. Our graphs of parameters versus environmental conditions and of state variables versus parameters do not tell us how fast things are changing (Figure 8.1). Temporal changes in environmental conditions, parameters, and state variables can be slow or rapid, persistent or intermittent. Three well-known dichotomies for temporal changes are persistent or brief (e.g. press versus pulse perturbations, Bender et al. 1984), continuous or jump-like (e.g. state threshold versus driver threshold, Andersen et al. 2009), and linear or step-like (e.g. ramp disturbance versus press disturbance, Lake 2000).

There have also been several attempts to link temporal changes in parameters to temporal changes in the ecosystem state or equilibrium state itself (Andersen et al. 2009, Lake 2000, Scheffer and Carpenter 2003). Lake (2000) notes that that environmental conditions (i.e. parameters) can change linearly over time, which he calls ramp disturbances. He goes on to point out that ramp disturbances need not cause gradual changes in equilibrium state and can give rise to steep but smooth threshold shifts in the equilibrium state. He labels this sort of sudden shift as a press response. Similarly, Andersen et al. (2009, their Figure 1) list three classes of thresholds—driver thresholds, state thresholds, and driver–state hysteresis. Driver thresholds are related to thresholds in parameters while state thresholds and driver–state hysteresis involve state variables themselves. Andersen et al.'s classification covers much of the same ground as Lake (2000) except for their inclusion of hysteresis. Finally, Scheffer and Carpenter (2003) distinguish between smooth, abrupt, and discontinuous regime shifts and they specifically link sudden shifts in state variables to discontinuous jumps.

It is often assumed or stated that the existence of a sudden and unexpected shift in either a parameter or state variable is prima facie evidence of multiple stable states (Beisner et al. 2003, deYoung et al. 2004, Knowlton 1992, Paine and Trimble 2004, Petraitis and Latham 1999) even though neither temporal changes in parameters nor transient behavior of state variables provide insights into existence of multiple stable states. Part of the problem is that the pace of change need not match the change in equilibrium state. For example, a state variable may approach a new equilibrium very gradually even if there has been a discontinuous jump from one stable point to another. State variables may move very slowly towards the new alternative equilibrium point because of demographic inertia or other life-history characteristics of the species involved. This point has been made early and often by both experimentalists and theoreticians (Connell and Sousa 1983, Frank 1968, Hastings 2004). The pattern of change over time tells us little or nothing about the possible states at equilibrium.

There is also no reason to assume that the pace of temporal changes between states is similar in both directions. The shift from ecosystem state A to state B could be gradual but then the shift from B to A could be rapid or threshold-like. The pace of the shifts depends on how parameters vary and on the biology of the species involved. For example the shift from grassland to forest could be gradual while the shift from forest to grassland could be quite rapid because trees grow more slowly and take longer to mature compared with grasses. There could even be smooth change in one direction and a discontinuous jump in the other. This is the pattern seen in one-jump paths (see Figure 6.5).

Demonstrations of thresholds in parameters or state variables provide little or no insight into the existence of multiple stable states. Yet the idea persists. For example, the loss of ecosystem resilience is often linked to sudden shifts in ecosystem states. Ecosystem resilience has been defined as the capacity of an ecosystem to resist external disturbances before shifting to a new state, and ecosystems that remain in a particular state in the face of large disturbances are viewed as being resilient (Holling 1973). Loss of resilience is seen as a "sudden, sharp, and dramatic change in the responding state variable" (Folke et al. 2004) and is commonly thought of as being a sudden change over time (i.e. the steep step-like change show in Figure 8.1e). The most common view is that a sudden shift in ecosystem state due to loss of resilience is caused by a shift between alternative states via a discontinuous jump (e.g. Folke et al. 2004; see their Figure 2). Yet it is just as plausible that the approach to an alternative state takes a very long time after a discontinuous jump. We need to remember that the jump refers to the equilibrium state and not the dynamics of the approach, and so the rapidity of change tells us little about the existence of multiple stable states.

8.1.1 Multiple stable states without thresholds

To emphasize the fact that multiple stable states do not require thresholds, Petraitis and Hoffman (2010) developed a grazing model with multiple stable states with no thresholds. They assumed the rate of change of the number of individuals in the exploited species, $\mathrm{d}N/\mathrm{d}t$, to be

$$\frac{\mathrm{d}N}{\mathrm{d}t} = s + Nf(N) - Nv(N) - pN \tag{8.1}$$

where s is the input supply of recruits that is unrelated to density, $f(N)$ is the per-capita effect of facilitation of recruitment of new individuals by established individuals in the population (e.g. enhancement of seed arrival and survival by the presence of adults), $v(N)$ is the per-capita mortality rate, and p is the rate of removal of adults by grazing. Per-capita recruitment is then the combined effect of facilitation and a mass supply rate. They assumed the per-capita effects of facilitation and mortality are nonlinear but did not have thresholds:

$$f(N) = aN + bN^2 \quad v(N) = c + dN^2. \tag{8.2}$$

The combined effect of recruitment and mortality without grazing (i.e. $p = 0$) is a cubic equation, and the rate of change in the population initially declines before giving rise to the familiar hump-shaped curve usually associated with logistic growth. There is an Allee effect due to the presence of adults. Adults cause enhancement of recruitment over and above the mass supply rate of recruits. The enhancement is a per-capita effect; the more adults present, then the larger the effect on the per-individual recruitment rate, although the additional benefit declines with density and the curve saturates (Figure 8.2). The individual contributions of mortality, mass supply, and facilitation are nonlinear but none show a threshold. When grazing is added, the equilibrium density of the exploited species is a balance between the population's rate of growth and the rate of grazing (Figure 8.2b).

The system has a cusp catastrophe even though none of the per-capita rates show threshold-like shifts with changes in density. As grazing pressure (p) increases, the system moves from one stable equilibrium point to two stable and one unstable point, and finally back to one stable point. The transition from one stable point to three points is discontinuous and occurs over a very small change in p. Thresholds and multiple stable states do exist in ecosystems, but the presence of a threshold in a parameter is not necessary for the existence of multiple stable states.

8.2 The state-and-transition concept implies models with multiple stable states

The state-and-transition concept has been the basis of virtually all models of multiple stable states in rangelands. We have already seen how some ecologists have attempted to link state-and-transition models to a cusp catastrophe (Rietkerk et al. 1996). Models based on the state-and-transition concept or state-and-threshold model usually include thresholds, irreversible shifts, and multiple states (Friedel 1991, George et al. 1992, Laycock 1991, Westoby et al. 1989), and thus are often called state-and-threshold models. Early state-and-transition models were presented as graphs or flow diagrams with boxes representing states and causal arrows connecting the boxes representing transitions (e.g. Laycock 1991).

Different states do not need to be multiple stable states within the state-and-transition formulation. Abrupt changes could easily be due to smooth threshold-like behavior (e.g. Figure 8.1c), and difficulty in reversing course could reflect slow responses rather than hysteresis. State-and-transition models are also often thought to represent a system not at equilibrium. This view was unfortunately reinforced by the title of Westoby et al.'s (1989) paper, "Opportunistic management for rangelands not at equilibrium." Cingolani et al. (2005), however, correctly note that most early proponents did

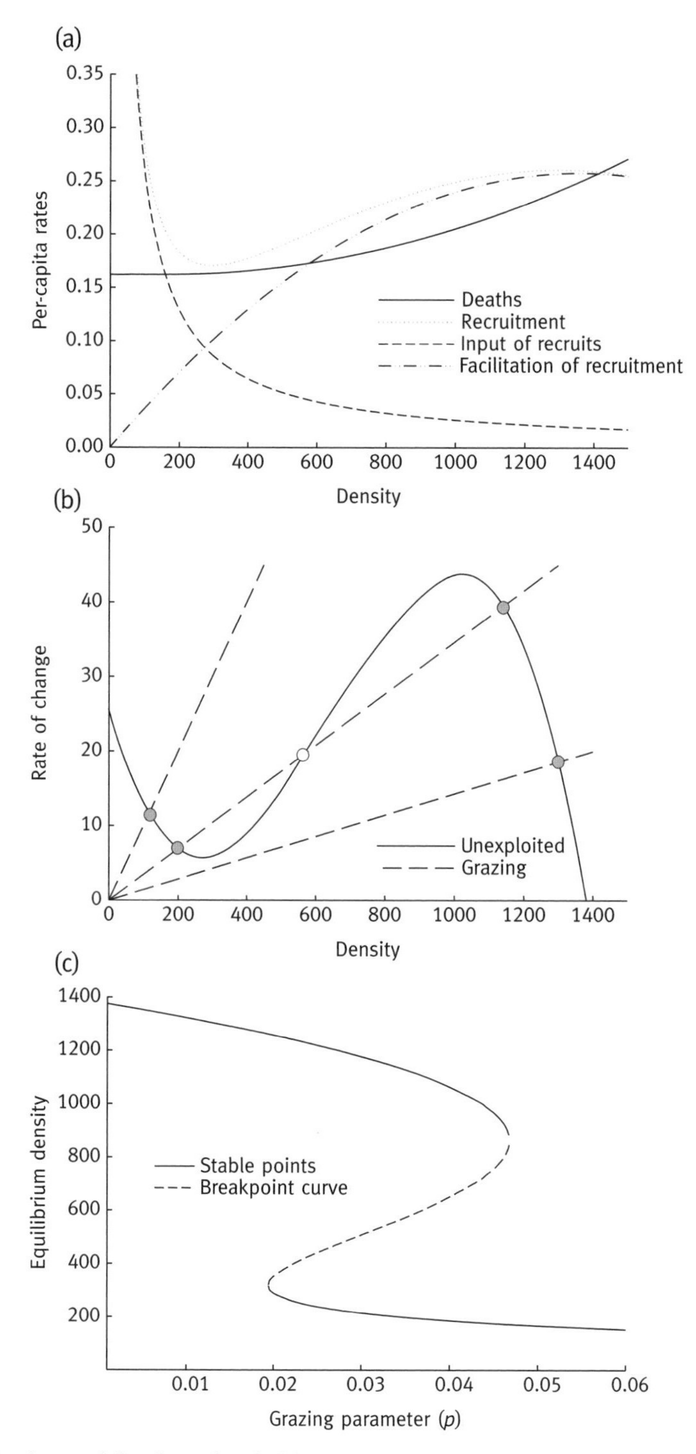

Figure 8.2 Grazing model without thresholds in parameters but containing multiple stable states. (a) The per-capita rates for the supply of recruits, facilitation, and mortality as functions of density. (b) The population growth rate in the absence of grazing (solid line, unexploited) and the harvest rates under three levels of grazing (dashed lines). The solid circles show stable equilibrium points and the open circle shows an unstable point. (c) The equilibrium densities as a function of the rate of grazing. Redrawn from Petraitis, P. S. and Hoffman, C. (2010). Multiple stable states and relationship between thresholds in processes and states. *Marine Ecology–Progress Series*, 413, 189–200. Reproduced with permission from Inter-Research Science Center.

not equate state-and-transition models with non-equilibrium systems as claimed in a number of reviews (e.g. Briske et al. 2003, Ellis and Swift 1988, Illius and O'Connor 1999).

Three observations about state-and-transition models are relevant to the theory of multiple stable states (Jackson and Allen-Diaz 2002). First, state-and-transition models are a way to organize observations, interviews, and data into a conceptual framework. The developers of state-and-transition models did not view them as a new and formal theory of vegetation change. In fact, Westoby et al. (1989) stated, "We are proposing the state-and-transition formulation because it is a practicable way to organize information for management, not because it follows from theoretical models about dynamics." Secondly, the state-and-transition framework can be used to "model" multiple states or a single climax state. The models based on this framework can be used to examine transitory dynamics (i.e. nonequilibrium models) or stable states (i.e. states at equilibrium). The boxes and arrows that are so often part of state-and-transition models can be written as a series of difference equations or in matrix form where the "states" are state variables and the transitions are parameters. The models can be deterministic or stochastic. Regardless of the formulation, state-and-transition models can have one stable state or two or more stable states. Finally, the state-and-transition framework does not require thresholds in parameters or state variables.

Jackson and Allen-Diaz (2002) voice the opinion that state-and-transition models are "heuristic, empirical tools that are flexible and general—and hence, powerful." We would counter that overly flexible and general models often do not provide specific outcomes that can be easily falsified. Moreover unusual and counter-intuitive observations are often explained away by making changes in the model. We would suggest that good models must provide clear and testable alternative hypotheses, and we will return to this point in Chapter 10.

8.3 The cup is defined by parameters and the ball by state variables

The cup and ball concept is a metaphor that helps us visualize multiple stable states. The hills and valleys are thought to be defined completely by the parameters, and the position of the ball provides information about the equilibrium state and its stability. The hills or ridges between basins of attraction are often thought to show the position of the unstable equilibrium point that always lies between two stable points. This need not be the case, as can be easily seen by using the ball and cup analogy to examine the dynamics of the Lotka–Volterra competition model for two species with a saddle node (Figure 8.3). Two stable points are each species by itself with a saddle node in between. In this system each species can occur by itself as a stable state. When either species is at its equilibrium point, it cannot be invaded by the other. There is also an unstable equilibrium point at the saddle node. At this point, any perturbation of the system, which shifts densities away from the saddle node, will cause the system to roll into one of the two basins of attraction (i.e. the cups) and move towards one of the stable points.

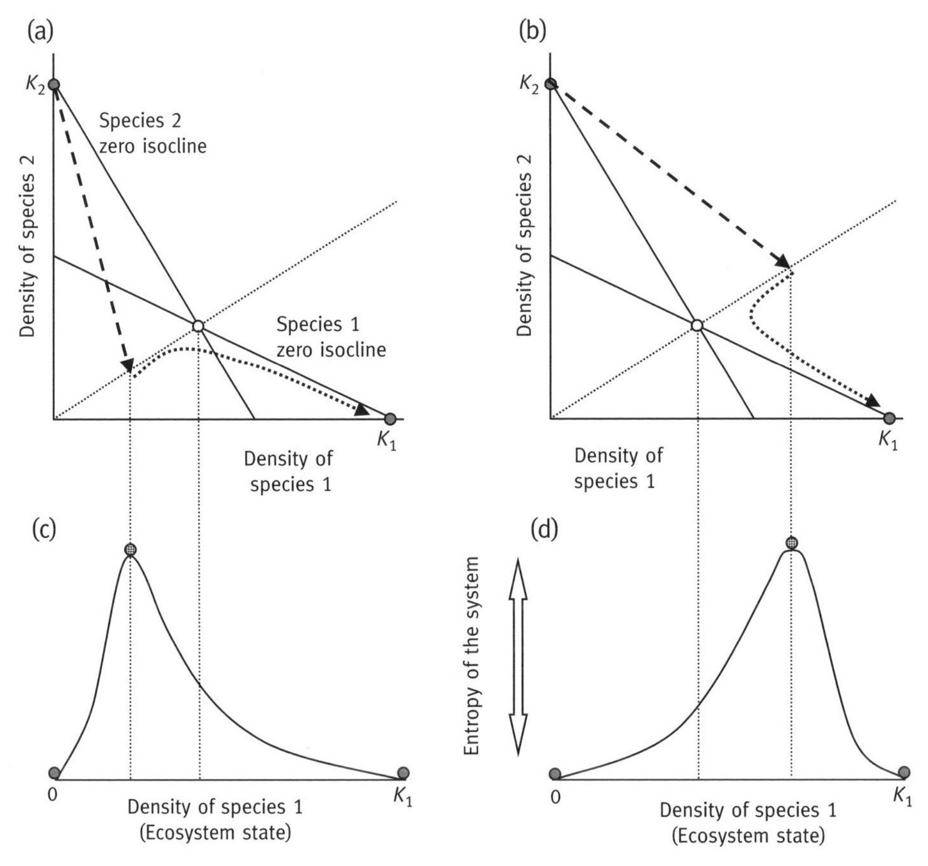

Figure 8.3 Comparisons between phase diagrams and the hill and valley representations of stability. (a) Phase diagram of a saddle node in the standard Lotka–Volterra model for two-species competition. Solid lines are zero net isoclines and the small dotted line is the separatrix. The dashed line is a pulse perturbation of species density across the separatrix and the large dotted line is the path towards the new equilibrium point after the perturbation. Open circles are unstable equilibrium points and closed circles are stable equilibrium points. (b) Same as in (a) but with a different perturbation. (c) The perturbation and separatrix from the phase diagram shown mapped onto the hill and valley representation of stability. (d) Mapping of (b). Redrawn from Petraitis, P. S. and Hoffman, C. (2010). Multiple stable states and relationship between thresholds in processes and states. *Marine Ecology–Progress Series*, 413, 189–200. Reproduced with permission from Inter-Research Science Center.

The boundary between the two basins of attraction is known as the separatrix (Slobodkin 1961).

The position of the ridge depends on how the combined densities of the two species are perturbed relative to the separatrix. We can see this by examining two different paths across the separatrix. For the first case, imagine that the perturbation of the density of species 2 is large relative to the perturbation of species 1 (Figure 8.3a), and for the second case, imagine the perturbation of both species is similar (Figure 8.3b). The position of the ridge that separates the two basins of attraction depends on where the separatrix is crossed, and so the position of the ridge shifts from one example to the other (Figure 8.3c versus 8.3d).

There are three crucial points. First, neither the separatrix nor the ridge in a cup and ball diagram defines the "position" of the unstable equilibrium state. The unstable equilibrium point—the saddle node—lies on the separatrix, and the ridge in a ball and cup diagram shows the position of the separatrix. A particular ball and cup diagram is but one possible slice of the phase space and that particular slice may not include the unstable equilibrium point. Second, the units on the axis showing the state of the ecosystem (e.g. the x-axis in Figure 8.3c) depend on how the slice is taken through the phase diagram (e.g. Figure 8.3a). Here we arbitrarily set the density of species 1 as the ecosystem state of interest, but the slice could just as easily have been drawn as a diagonal line connecting K_1 and K_2. If this were the case, the units of the ecosystem state would be a combination of the densities of species 1 and 2.

Third, the contour of the ridge and valleys does not indicate the direction of change in ecosystem state. For example in Figure 8.3(b), the density of species 1 first declines after the perturbation before increasing to K_1, and so the slope in Figure 8.3(d) provides no information about changes in density of species 1 once the separatrix is crossed. The simple representation of a single ridge hides the many alternative paths across the separatrix from one basin of attraction to another. Any single representation is only one of many possible snapshots of the positions of the stable points and the separatrix. The representation of stability as a ball on a surface is little more than a cartoon and may not always provide us with the correct intuition.

The cup and ball concept has also contributed to the confusion about the link between thresholds and multiple stable states. Part of the problem is that there are two definitions of thresholds. May (1977) in his discussion of multiple stable states used the term threshold in a much narrower sense than is used in most current discussions. For May, thresholds are discontinuous jumps in the equilibrium state of variables in response to small, continuous, and smooth changes in a parameter. In contrast, ecologists often use the term threshold for the ridge between basins of attraction.

8.4 Systems with multiple stable states usually involve specific features of species or the environment

It has been suggested that multiple stable states are associated with ecosystem engineers, positive feedback loops, and stressful abiotic conditions (Didham et al. 2005, Knowlton 1992, 2004, Petraitis and Latham 1999, Wilson and Agnew 1992). However, all of these ecological features can be found in systems that lack multiple stable states. Conversely, systems with multiple stable states can occur without the presence of ecosystem engineers, feedbacks, or harsh environmental conditions. While certain characteristics of species and the environment may increase the possibility of multiple stable states, the presence of ecosystem engineers or specific environmental conditions does not guarantee the existence of multiple stable states.

The linkage from characteristics of species and the environment to multiple stable states is often made by invoking a threshold. Ecological engineers, positive feedback loops, and environmental conditions often cause threshold-like changes in environmental conditions. These threshold effects then in turn promote the occurrence of multiple stable states. We have already seen that thresholds are not required for multiple stable states, and so it should be no surprise that particular features of the environment or species are also not necessary.

Even so, this misconception is quite evident in descriptions and models of multiple stable states involving positive feedbacks. The widely cited review by Wilson and Agnew (1992) of positive feedbacks in plant communities has been very influential in the persistence of this notion. They suggested that positive feedback switches are important in promoting spatial and temporal patterns among different plant communities. They identified four types of switches and argued that these switches are responsible for spatial mosaics, sharp boundaries, and distinct changes in the pace of succession. A switch is a positive feedback loop in which a vegetative state modifies and reinforces the local environmental conditions so that the persistence of that state is favored. There may also be inhibition of an alternative vegetative state. Wilson and Agnew summarized many examples of switches that are mediated by abiotic factors such as fire, water, wind, light, and pH, and by biotic factors such as herbivores and soil microbes. It is important to note that they used the term "switch" for the positive feedbacks that maintain states and not for the shift between alternative states. They used conceptual models and graphs to support the idea that the shift between alternative states is threshold-like. Wilson and Agnew are still often cited in discussions of threshold, multiple stable states, and state-and-transition models (e.g. Briske et al. 2008, 2010, Eppinga et al. 2009, Hotes et al. 2010, Kitzberger et al. 2012, Odion et al. 2010).

Wilson and Agnew (1992) asserted that positive feedback switches will give rise to multiple stable states, but provided very little evidence to support this view. Many of their examples appear to have multiple stable states and positive feedback switches. It is completely plausible that the two are linked. Wilson and Agnew, however, mined the literature for examples of positive feedback switches and of strong spatial and temporal patterning. It should not be surprising that they found examples of switches and that these switches appeared to drive multiple stable states. It may be interesting to mine the literature for examples of multiple stable states without positive feedback switches or examples of switches without multiple states. This may be difficult to do since the two are so closely linked in the minds of most ecologists.

There are many models of multiple stable states that provide counter-examples. The simplest and clearest case would be existence of two stable states in the Lotka–Volterra model of competition. This is a model based on purely competitive effects—in other words, negative feedbacks. Each species adversely affects its competitor and itself. There can be two stable states without the presence of the switches envisioned by

Wilson and Agnew. The reverse, however, may be true. Positive feedback switches may enhance the development and maintenance of multiple stable states. How common this may be remains an open question.

In contrast, Didham et al. (2005) suggested that multiple stable states are more likely to occur under harsh environmental conditions (however, see Mason et al. 2007). Their argument relies on evolutionary and ecological consequences for species that live in environmentally stressful habitats. On the evolutionary side, species in stressful environments tend to show convergence of traits. For example, sclerophyll scrubs living in Mediterranean-type climates, which have wet mild winters and hot dry summers, show similar adaptations. The convergence of traits and the similar aspect of the plant communities found in the Mediterranean Basin with the plant traits and communities found in the Californian chaparral, the Chilean matorral, the South African fynbos, and the Australian mallee is quite stunning (Dallman 1998). This convergence of traits is thought to have given rise to similar patterns of life history, thus resulting in species with nearly equal competitive effects, an effect known as "trait under-dispersion" in the current plant literature (Weiher and Keddy 1995).

Evolutionary convergence of plant traits under harsh environmental conditions has three ecological consequences (Didham et al. 2005): few successful offspring (known as propagule limitation), stochastic priority effects, and variation in species pools regionally. Didham et al. argued that these three combine to give rise to alternative states. Propagule limitation can be due to a limited supply of viable seed and/or short dispersal distances and this contributes to uncertainty in the order of arrival of colonizers. Stochastic priority effects simply mean that a species that arrives early cannot be displaced by species that arrive later. The outcome is "stochastic" or random due to propagule limitation and variation in the regional species pool.

Didham et al.'s scenario is not easily translated into a model of multiple stable states. If their line of reasoning were to be formalized into a model, it would be more akin to a Markovian process or a mutation-drift model in which the starting point is random and the walk to fixation of one or another state is a random walk. As we discussed in Section 7.3, while these models can give rise to alternative states the models and the dynamics of these models are quite different from the conventional notion of multiple stable states as encapsulated by catastrophe theory. Didham et al.'s scenario has more in common with Hubbell's (2001) neutral theory of biodiversity.

There are also lines of contradictory evidence that must be considered. First, there are clear examples from benign environments of ecosystems with multiple stable states—for example, tropical woodlands and coral reefs (e.g. Dublin et al. 1990, Norström et al. 2009). Knowlton (1992) argued that coral reefs are likely to have multiple stable states because "they are not subject to routine, seasonal resetting from strong annual changes in the physical environment." Chase (2003b) also suggested that multiple stable states and the importance of history are more likely in systems with large regional species pools, low rates of connectance, high productivity, and low disturbance. High productivity and low disturbance are characteristics of benign systems and so more species have a chance. Chase's idea is very similar to Sale's ideas of lottery models (Sale 1977) and covers much of the same ground as DeAngelis and Waterhouse (1987), although they used different terminology. Similar ideas can be found in Drake's work (Drake 1990, 1991, Drake et al. 1993, 1994, 1996)

Some researchers suggest that ecosystem engineers enhance the likelihood of multiple stable states. For example, using a metapopulation model, Shurin et al. (2004) found that multiple stable states can occur through habitat modification by ecosystem engineers (what they called "biotic heterogeneity") although the system required some kind of abiotic heterogeneity to maintain multiple stable states. Yet ecosystem engineers are the paradigm of species with exceptional impacts on communities, and they represent the exact opposite of Didham et al.'s position that communities are more likely to have multiple states if they contain species with near equal competitive effects. Finally there are many communities in harsh environments and with trait under-dispersion that do not show any signs of multiple stable states. The convergent communities of sclerophyll scrubs living in Mediterranean-type climates are a striking example.

Much of the difficulty of equating the likelihood of multiple stable states with particular characteristics of the environment and species comes about during the translation of our perceptions about ecological processes into the parameters and state variables of models. Ecologists often think of environmental conditions as drivers of changes in ecosystems. These conditions or drivers are then translated into model parameters. Yet the translation can be imprecise or incomplete.

This is most easily seen if we start with a model and attempt to match model parameters with environmental conditions. Suppose we have the typical notched pattern that is seen in a cusp catastrophe (Figure 8.4). The model has two parameters, a and b, and the notch is oriented so it runs parallel to the axis for parameter b. Now we usually assume there is one to one matching of parameters to environmental conditions and other drivers. For example, in Rietkerk et al.'s (1996) model of grazing in the Sahel, rainfall is equated with parameter b and grazing with a (Figure 5.1). However, it is equally likely that rainfall and grazing are not orthogonal with b and a, respectively. The axes of rainfall and grazing also may not be orthogonal to each other when mapped in parameter space. Finally, there may be scaling differences; for example, the effects of grazing could be multiplicative and thus on a log-scale while the effects of rainfall could be additive. As a result, the notch pattern would be rotated and distorted when plotted in terms of grazing and rainfall (Figure 8.4). Loehle's (1989b) graph of a butterfly catastrophe is an example in which pattern of the cusps and folds are distorted and not orthogonal to the axes of grazing and rainfall (see Figure 5.6). The effects of rotation and distortion will depend in part on how we perceive environmental conditions. We usually think of environmental conditions as been independent—that is their axes run at right angles to each other. However, in many cases environmental conditions are not independent, and when this occurs we may end up with a model that does not match our observations from nature.

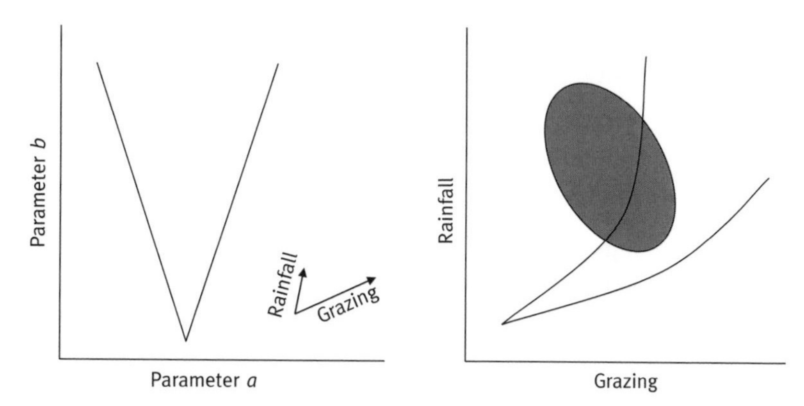

Figure 8.4 The mismatch between parameters and environmental conditions. The left-hand panel shows how environmental conditions such as rainfall and grazing may not be independent (i.e. the vectors are not perpendicular), and how parameters may not match environmental conditions. The right-hand panel shows how actual environmental conditions (the shaded area) may not encompass the entire region of bifurcation.

It is also possible that not all combinations of environmental conditions occur in nature (Figure 8.4). Summer rainfall and summer temperatures covary in semi-arid regions of the south-western United States. Drought conditions are usually accompanied by very warm weather. As a result only a small portion of the parameter space would be found in nature and would give rise to a truncated graph (e.g. Figure 3.5).

The end result is that a fit between observations about environmental conditions and model parameters may be incomplete or imprecise. This can lead to the assumption of particular characteristics of the environment and/or of the species involved being necessary for the existence of multiple stable states. This problem is not unique to models of multiple stable states, but it can troubling given that models of multiple stable states can give rise to very complex behavior as parameters vary. If there is a mismatch between our preconceptions about environmental conditions and model parameters we may end up misinterpreting the underlying dynamics of the ecosystem. It is not unlike the problem of choosing the most appropriate map, which we discussed in Chapter 1.

8.5 Concluding remarks

All of the misconceptions about multiple stable states that we have discussed are easily found throughout the literature. The notions that multiple stable states involve systems that always show hysteresis, are nonequilibrial, and require stochastic processes are probably due to an incomplete understanding of the underlying theory. In much the same way, the under-appreciation of what theory can tell us has led to the unfortunate linkage of state-and-transition models with multiple stable states and to the over-reliance on the cup and ball analog to describe the dynamics of systems with multiple stable states. Multiple stable states also do not require thresholds or particular aspects of species and of the environment. The development of models depends on making plausible links between natural phenomena and parameters, and most of the misconceptions about what theory can tell us about multiple stable states arise from mistranslations of nature into parameters.

9

Using temporal and spatial patterns as evidence

Attempting to infer ecological processes from patterns in nature is a dangerous business. No one—not even scientists—approaches patterns in everyday life without preconceived notions about the underlying causes and what can be deduced from observations. Given that we have preconceptions, observations of patterns in nature are not unbiased or neutral in meaning. Darwin wrote to Henry Fawcett in 1861 that "there has been talk that geologists ought only to observe and not theorize . . . at this rate a man might as well go into a gravel-pit and count the pebbles and describe the colors. How odd it is that anyone should not see that all observation must be for or against some view if it is to be of any service!" (quote from Gould 1992). While preconceptions can easily mislead us, it is our ideas about causality that allow us to decide if an observation is for or against our view. In the best of all possible worlds we deal with our biases via experimental manipulation, which relies on replication, randomization, and controls. Yet in many cases the temporal and spatial scales of ecological processes are too long and too broad for us to undertake informative experiments. The more cynical view is that it is cheaper, easier, and quicker to collect observational data than to undertake a well-replicated large-scale experiment with appropriate controls and randomization. Regardless of the motivation, patterns in nature are often used by ecologists to infer processes and causality.

Even so, description of patterns is, by far, the most common way by which ecologists attempt to find evidence for multiple stable states. There are many critiques of why this sort of approach is unproductive and untenable (Underwood 1991), but we will discuss the problems and issues that arise from trying to infer the existence of multiple stable states from observations of patterns in time and space. We will provide a few examples of the kinds of patterns that have been used, but we will not attempt to catalog the large number of studies that have been put forth as evidence since they offer little in the way of proof.

9.1 Inferring causality from patterns

Using patterns to detect ecological processes presents several serious challenges. The first problem is contextual. As Darwin noted, observations are only valuable if they are for or against an idea, so we must first define our viewpoint. For ecologists, this usually entails creating a model. The model may be conceptual or functional, but it provides the framework for our observations. Second, we must assume that the proposed model is, in fact, an accurate representation of the ecological processes operating in the ecosystem under observation. For example, it is quite easy to write a computer program to mimic on the screen the movement of the turning of the hands and the swing of the pendulum of a grandfather clock, but the code will not capture the reality of the gears behind the face of the clock. Simply making a model that matches an observed pattern in nature does not mean the model provides the correct explanation.

The third challenge is how we determine whether an observed pattern in nature actually matches the model. Confirmation of a model should be carried out with field observations that were not used to develop the model. While this should be obvious, it is rarely done. To the best of our knowledge, no one using models to evaluate observational data for evidence of multiple stable states has split the data and used half for model development and the other half for testing model goodness-of-fit. It is likely that nearly all "tests" for multiple stable states based on consistency between patterns in nature and models in hand rely on models that were developed post-hoc and with prior knowledge of the pattern. Using the same data for model development and for confirmation of patterns in nature reminds one of the old adage about which came first, the chicken or the egg.

Finally there is the question of how close is close enough when matching patterns to models. We may be unable to distinguish a difference between an observed pattern and a proposed model simply because we do not have enough data. Thus the agreement between the model and the pattern may reflect low statistical power, which leads to the classic mistake of accepting the null hypothesis. Failure to reject a hypothesis does not imply that it is true; it simply means we lack the ability to distinguish. This then begs the question of how much data do we really need and when is the match good enough. One approach that addresses this problem, and is gaining some ground among ecologists, is tests of bioequivalence (Cole and McBride 2004, Dixon and Pechmann 2005, MacKenzie and Kendall 2002, Mapstone 1995, Robinson et al. 2005, Romano 2005). Tests of bioequivalence are commonly used by governmental regulatory agencies and in the medical profession to answer questions of drug safety. The question, "Is drug X safe?" implies safety in terms of a particular risk (e.g. does drug X cause cancer?) and knowledge of the normal bounds of safety beforehand (what the population-wide rates of cancer are). This approach has not been used to evaluate models of multiple stable states against observed patterns, but it would require identification of a phenomenon associated with multiple stable states, for example discontinuous jumps, a means of measuring it, knowledge of normal variation, and assessment of risk if the wrong conclusion is reached. These requirements suggest that a well-designed test based on observational data would not be cheap, easy, or quick.

An alternative approach would be to use competing models and use measures of relative support to determine the best model (Angilletta et al. 2006, Burnham and

Anderson 2002). This is potentially a very powerful approach but it has not been used by modelers of multiple stable states. The logic would be to develop models with and without multiple stable states and determine which model is the best description of an observed pattern. This, in fact, is the basis of Ditlevsen and Johnsen's (2010) critique of Dakos et al.'s (2008) statements about critical slowing down. Ditlevsen and Johnson showed that an alternative model provided a better explanation for the pattern.

Given the tight link between observation of patterns and models, our discussion of patterns must be tied to models. For multiple stable states, the modeling approach varies with the area of ecological research. Observational data from studies of rangelands, grasslands, and semi-arid ecosystems tend to be linked to state-and-transition models. Analyses of patterns in forests and coral reefs are usually simulations based on occupancy models because individuals can be identified, and more importantly the persistence of an individual tree or coral colony is entirely dependent on occupancy of a specific site. While there are exceptions, studies of spatial and temporal patterns in nearly all other ecosystems rely on the classic models that are based on differential equations. Regardless of the type of model used, all rely on detection of one of the flags of catastrophes.

9.2 Linking patterns to catastrophe flags

Spatial and temporal patterns that are usually assumed to be indicative of multiple stable states can be related to catastrophe flags. These connections, however, are rarely acknowledged by ecological researchers. The flags of particular interest are hysteresis, divergence, sudden discontinuous jumps, anomalous variances, inaccessibility, and modality. Inaccessibility and modality are structural and are the essence of the idea that multiple stable states involve the occurrence of two or more states at the same time or place. Sudden discontinuous jumps and divergence are the result of these structural features. We also must keep in mind that the occurrence of hysteresis depends on the delay convention, and that the occurrence of anomalous variances and the related phenomenon of critical slowing down are not necessarily associated with catastrophes and multiple stable states.

Several critical assumptions must be made in order to link these flags to spatial and temporal patterns. Let us first consider two assumptions needed to link hysteresis, divergence, sudden jumps, and anomalous variance to patterns. First, and most importantly, we must assume some connection between changes in parameters and changes over time and space. The strict definitions of these flags of catastrophe require the observed changes in state variables to be driven by changes in parameters (see Figures 6.5 and 6.6). There is no mention of time or space in the definitions. In order to infer the occurrence of these flags from spatial and temporal patterns, we must assume that changing the condition of a system over time or across space is the same as changing or tuning a parameter. There are a few exceptions, and those involve cases where an environmental condition that is plausibly related to a parameter is tracked over space or time. For example, observations of patterns in rangelands occasionally include descriptions of changes in the amounts of rainfall and grazing or the frequency of fire. In these situations, a parameter—for example, grazing—is observed to change. Even so, we are unaware of any good examples in which both state variables and parameters are monitored with adequate levels of replication and reasonable experimental controls.

The second critical assumption depends on linking discontinuous jumps to the tempo of transient dynamics. We know in systems with multiple stable states that there will be a discontinuous jump from one equilibrium state to another as parameters change and the fold of a catastrophe is passed. It is, however, a statement about the potential of the system and not the rapidity with which the new equilibrium point is approached (e.g. see Figure 8.1). While observations of abrupt shifts in state variables over time or space are often linked to a discontinuous jump, this need not be the case. Crossing a fold and having a discontinuous jump does not necessarily give rise to a rapid shift over time or space. The response in state variables may be quite slow or diffuse and does not need to look like a sharp threshold. Moreover it is likely to be very difficult to distinguish between a smooth but steep shift associated with a phase shift and a discontinuous jump that arises from the typical S-shaped curve of a cusp catastrophe. The rapidity of response provides no insight, and so we must assume that rapidity is related to discontinuous jumps.

There is also a confusing use of the term “divergence.” Divergence as a flag of catastrophe refers to ending up at different equilibrium states when there are small differences in initial parameter conditions (see Figure 6.5). The small difference in parameters means there will be a small difference in equilibrium states when the system occupies a part of the parameter space which is “smooth” (i.e. there are no folds or cusps nearby). Divergence as a flag occurs when a parameter is then changed and the two equilibrium points diverge. One goes “uphill” on the top of the S-shaped curve and the other goes “downhill” on the bottom arm of the curve as a parameter changes. There is, again, no concept of changes over time or space—only changes in equilibrium states as parameters change. Yet ecologists by and large take divergence to mean divergence after a perturbation of state variables or differences in starting conditions of the state variables. For example, for their test of divergence, Schröder et al. (2005, in their Figure 2) state “different initial values of the state variable... lead to different stable states, when situated in different basins of attraction.” This sort of divergence is due to modality and inaccessibility. It is related to the flag of divergence only if there is a matching between changes in parameters and spatial or temporal change.

Spatial and temporal patterns that are used as evidence for multiple stable states are also quite different in scale. Observations taken over time tend to come from long-term studies done in a single place. An extreme example of this is the analysis of a 250-year time series of pollen based on five sediment cores from one lake (Sayer et al. 2010). In contrast, observational data from spatial surveys are often collected from a large number of sites but with each site being sampled over a relatively short time interval. For example, Schallenberg and Sorrell (2009) examined the shifts between a clear-water state and a turbid-water state in 95 New Zealand lakes, with each lake being sampled only briefly. Temporal data are akin to a movie shot at a single place and spatial data are more like a collection of single snapshots of many places.

Despite these difficulties, there have been a number of suggested tests for multiple stable states that can be categorized by catastrophe flags (Table 9.1). These tests by and large are not detailed protocols for experimental designs but rather descriptions of spatial and temporal patterns that could plausibly arise if a system contained multiple stable states. To some extent all require detection of sudden changes in state variables. This can be shifts over time or space (IIIa and IIIb, respectively, in Table 9.1), perturbation of parameters or state variables (IIIc through IIIf) or changes in variance (IIIg). In all of these tests, the effects of environmental conditions, parameters, and state variables are confounded with changes over time and space. Moreover, Scheffer and

Table 9.1 Patterns over space and time that have been suggested as tests of multiple stable states organized by catastrophe flags. Proposed tests contain either smooth changes in parameters and/or perturbations of state variables. Perturbations of state variables include cases in which the starting conditions of the state variables differ.

Catastrophe flags	Continuous smooth change in parameters over time or space	Pulse perturbation in state variable	Proposed tests or patterns[a]
Modality and inaccessibility	No	Yes	I(a), II(b), II(c), II (d), III(e), IV(b)
Hysteresis	Yes	No	II(a), III(f)
Jumps	Yes	No	I(b), III(a), III(b), IV(a)
Anomalous variance	Yes	No	III(g)
Divergence	Yes	Yes	III(d)

[a] Codes for proposed tests or patterns; note that some codes appear in more than one row:

I. Beisner et al. (2003, Figure 1): a, perturbation of state variables; b, changes in parameters.

II. Schröder et al.'s (2005) tests for multiple stable states: a, test for discontinuity; b, test for nonrecovery; c, test for divergence (with or without random variation in state variables).

III. Scheffer, Carpenter and co-workers' classification of patterns (Scheffer and Carpenter 2003, Carpenter and Brock 2006): a, a discrete step of state variables in a time series; b, a bimodal or multimodal distribution of state variables; c, different functional relationship in different basins of attraction (i.e. two different series of values for state variables over the same range of parameter values); d, state variables go to different stable states when starting conditions of state variables differ; e, state variables switch to an alternative state when state variables are perturbed; f, dynamics of state variables depend on whether the parameter is increased or decreased (i.e. there is hysteresis); g, second derivative of state variable over time has peaks (i.e. anomalous variance).

IV. Lenton's (2011) classification: a, bifurcations (changes in parameters near folds); b, noise-induced transitions (variation in state variables near folds).

Carpenter (2003) noted that these patterns can be linked not only to discontinuous jumps but also to smooth threshold-like changes.

Many of the proposed tests contain elements of Peterson's criteria. Schröder et al. (2005) for example define a test of nonrecovery as the detection of a failure to return to an initial equilibrium of state variables after a perturbation is removed. Similarly Lenton (2011) defines a noise-induced transition as a stochastic perturbation that pushes state variables across the breakpoint ridge between basins of attraction. Both tests imply that the same place is able to give rise to alternative stable states, which is functionally equivalent to some of Peterson's (1984) criteria. We should again emphasize these tests are based on the flags of modality and inaccessibility and not to the flag of divergence.

9.3 Biases in selecting and assessing evidence

Biases are always present when evidence is culled from the literature. We have already touched upon problems of preconceptions and of linking patterns to causality. Yet even if those issues can be overcome, we still face the problem of bias in our selection, assessment, and analysis of evidence. These biases are especially serious when we consider the extraordinarily large number of observational studies that have used spatial or temporal patterns as evidence for or against the existence of multiple stable states.

Biases in the selection of evidence tend to be due to overlooked or missed examples. In many cases the reason is benign; examples are missed because they lie outside the scope of a review or outside the area of expertise of the authors. Reviews of multiple stable states focused on specific ecosystems—for example open oceans, benthic marine systems, coral reefs, lakes, rangelands—share few examples in common (e.g. Cingolani et al. 2005, Dent et al. 2002, deYoung et al. 2004, 2008, Knowlton 1992, 2004, Lenton 2011, McManus and Polsenberg 2004, Norström et al. 2009, Petraitis and Dudgeon 2004, Scheffer et al. 1993). Even reviews with a broader focus overlook examples. For example, the 35 studies listed in the review by Schröder et al. (2005) do not include many of the examples given in other more narrow reviews. There can also be a bias due to the type of pattern being addressed. Lenton (2011) reviewed 12 ecological and nine climate studies for changes in temporal and spatial patterns associated with anomalous variance and skewness, and critical slowing down. Most are modeling efforts which demonstrate that anomalous variation and critical slowing down can arise in plausible ecological models with thresholds. Only nine of the 21 studies included data, of which three were lab experiments, one used field observations, and five used paleontological records. None of the nine studies is cited in other reviews of evidence for multiple stable states that focus on the other more familiar flags of catastrophes (e.g. Collie et al. 2004, Knowlton 2004, McManus and Polsenberg 2004, Schröder et al. 2005).

It is also difficult to escape biases due to lumping studies from different ecosystems or different approaches. Evidence in favor of multiple stable states is more likely to be found in lab studies rather than field studies. Schröder et al. (2005) reported that most of the supportive studies were done in the lab (10 out of 12) rather than in the field (three out of nine). Even within a particular ecosystem, there is a problem of comparing across studies and this can clearly be seen in comparisons among coral reefs.

Norström et al. (2009) reviewed 16 cases of coral–algal phase shifts throughout the world. Algae have replaced corals in many reef systems, often quite quickly and in a surprisingly complete fashion, and the shifts from coral to algae are not easily reversed. However, coral reefs differ from one part of the world to the next, and so it is not clear if shifts from corals to macroalgae are functionally equivalent. Just in terms of number of species, the reefs of the Indo-Pacific Ocean contain many more species than Caribbean reefs. There are also different kinds of coral reefs. While hard corals—scleractinians—form the backbone of many reef systems, soft corals, sponges, and corallimorpharians can be common members of some coral reef communities. Thus it is reasonable to ask if all reefs can be treated as the same. Clearly, reefs in different oceans are not the same but are they similar enough at some level that we can assume the shift from corals to algae to be functionally the same? This is an important consideration when discussing multiple stable states because of the underlying assumptions of modality and inaccessibility.

Assessment biases arise in a variety of ways but often revolve around use of different criteria for assessment. Studies rejected by one researcher may be included by others. Schröder et al. (2005) rejected 21 studies—many of which have been cited by others as good examples of multiple stable states. About 62% of the included studies show some evidence for multiple stable states. In contrast, Lenton (2011) reported finding changes consistent with critical slowing down, anomalous variance, and/or skewness in all but one study. The discrepancy between the two reviews is striking given the difficulty of detecting critical slowing down, anomalous variance, and skewness.

Assessment biases can also occur within a single study. For example, 37 of the 95 New Zealand lakes studied by Schallenberg and Sorrell (2009) showed shifts between clear-water and turbid-water states. The remaining 58 lakes showed no change. The authors found "regime shifts" were correlated with deforestation and the introduction of exotic fishes. However, the assignment to states of either clear or turbid was not done by a researcher unaware of the expected pattern, and the connection to deforestation and exotic fishes was found via data mining. Within sociological research this is known as agenda-driven bias. In ecology, the use of post-hoc assignments and uncontrolled manipulations has been fully discussed in critiques of "natural experiments" (Underwood 1990, 1991).

Biases in analysis involve how individual studies are counted. All current reviews of multiple stable states use "vote counting" tallies. Each study is assumed to have equal weight and evidence pro or con is evaluated as a vote. The shortcomings of this sort of approach have been widely discussed within the context of using meta-analysis in ecology (Englund et al. 1999, Gurevitch and Hedges 1999, Osenberg et al. 1999). The most serious bias is that studies in which no pattern is found (i.e. no significant result) are not published; this is known as the file drawer problem (Csada et al. 1996, Rosenthal 1979). Even a well-designed meta-analysis can still suffer from this problem.

There has not been a meta-analysis of experiments or patterns that have been cited as evidence of multiple stable states, and it is problematic whether such an analysis is possible. If Schröder et al.'s (2005) tally of acceptable studies is accurate, then there are far too few studies in hand for a meta-analysis. It is also not clear what would be the appropriate metric for measuring any of the flags of catastrophes as an effect size. Effect size for continuous variables is

$$d = \frac{\overline{T} - \overline{C}}{s} J(N)$$

where $\overline{T}$ and $\overline{C}$ are the averages from the treatment and control groups, s is an estimate of the pooled standard deviation, and $J(N)$ is a correction for sample size (Gurevitch and Hedges 1999, Petraitis 1998). The question then is how to conceive the averages as a measure of a catastrophe flag. For the detection of patterns, the averages could be a measure of state variables before and after a shift in state as seen in a time series or among sites that were classified into one of two states—for example clear and turbid lakes in Schallenberg and Sorrell's (2009) study. However, it is not clear how the difference between the two states is related to, say, modality or divergence. We could easily discover that the two states are different. We do not know if those different states occur in the same environment—that is under the same set of parameters. Finally, even if we could come up with a reasonable measure of effect size for one of the flags, there remains a problem of missed and overlooked studies. Recently it has been pointed out that effect size tends to diminish as experiments are repeated (known as the decline effect; Schooler 2011), and that publication bias tends to inflate the significance tests even within a well-designed meta-analysis (Francis 2012).

9.4 Evidence from temporal patterns

Patterns through time used in support of multiple stable states require long series of observations. A common type of dataset is a time-series derived from historical or paleontological sources, such as data on pollen or diatoms from cores of lake sediments. Collecting data for extant species or existing ecosystems is much more labor-intensive and is usually beyond the means of an individual researcher. Thus long-term time-series data are often are the result of collaborative efforts. A common example is the data that were collected from a species or an ecosystem of commercial importance and for which there is usually a governmental infrastructure to collect and maintain the data. Examples include annual catch data from commercial fisheries and changes in species composition in rangelands over time. Another example is censuses of species or ecosystems of particular interest to environmentalists or society—for example endangered or endemic species or specific habitats such as serpentine barrens. These data tend to be similar to data on commercial species although they tend to be shorter-term and often not maintained. Third are data from long-term monitoring at established field stations or long-term environmental research (LTER) sites. Data from site-specific and species-specific programs tend to be more ecologically rich but collected on a smaller spatial scale than data that are a byproduct of commercial enterprises.

9.4.1 Examples from marine systems

Catch data collected before and after collapses of commercial fisheries are often cited as evidence for multiple stable states and have prompted the development of models of multiple stable states. For example, Collie et al. (2004) used a model of multiple stable states to explain the collapse of George's Bank haddock (*Melanogrammus aeglefinus*). From the 1930s to the 1950s, haddock stocks were estimated to be 100,000–180,000 tons with a harvesting rate between 20 and 45% of estimated biomass. Haddock biomass spiked sharply to over 400,000 tons in the early 1960s, followed by a rapid crash that is believed to be triggered by high rates of harvesting (Fogarty and Murawski 1998). Biomass wavered below 50,000 tons for most of the remainder of the 20th century, with a brief moderate recovery peaking at 100,000 tons in 1980 and another recovery of similar magnitude leading up to the year 2000. With the exception of 3 years, harvesting rate after the 1960s was low (9–45%). Using these data and assuming there were two states—a haddock-rich state prior to the 1960s and a haddock-poor state after the 1960s—Collie et al. (2004) developed a predator–prey model with two thresholds and hysteresis in harvesting rate. With increasing harvesting rates, they found a rapid drop from the haddock-rich state to the haddock-poor state at a harvesting rate of 36%. Under declining rates, the model also contained a sharp transition from the haddock-poor state to the haddock-rich state at a harvesting rate of 21%. This hysteresis is consistent with the notion that haddock dynamics exhibit a discontinuous regime shift between alternative stable states.

Similarly, van Leeuwen et al. (2008) used data on catches of the Baltic Sea cod (*Gadus morhua*) and the collapse of the fishery to develop a model of multiple stable states. In the Baltic Sea, cod stocks and fishing pressure were high from 1974 to 1987. Between 1988 and 1993, cod biomass plummeted despite there being no obvious change in harvesting rates. It is assumed that climate forcing from 1988 to 1993 led to decreases in salinity and oxygen, and concurrent increases in temperature and nutrients. Environmental conditions reversed to previous levels between 1994 and 2005, but cod biomass remained low (Möllmann et al. 2009). Van Leeuwen et al. (2008) proposed that high and low cod biomasses were alternative stable states that were controlled by the interaction between adult cod and their prey, the juvenile sprat (*Sprattus sprattus*). Without cod present, sprat are common and face strong intraspecific competition which causes them to grow slowly and remain in a young adult phase for a long period. Young adult sprat are too large to be eaten by adult cod and not large enough to reproduce a sufficient number of juveniles to serve as prey for cod and thus promote cod recovery. Van Leeuwen et al. (2008) suggested that prior to 1988 adult cod were present in sufficient numbers to reduce the number of sprat and release them intraspecific competition. This release promotes rapid maturation and reproduction which in turn provides food for cod and supports high cod densities. Changes in salinity, oxygen, temperature, and nutrients caused the initial crash of cod stocks, and the resulting shifts in sprat demography prevented the recovery of cod. Van Leeuwen et al. called their scenario an "emergent Allee effect" because it is based on scarcity of adult cod.

There have been several studies done in benthic ecosystems looking at sudden changes over time. In the western Pacific a 25-year time series shows abrupt temporal changes in mussel and macroalgal cover in response to experimental removal of seastars (*Pisaster ochraceus*) and to El Niño (Paine 1974, Paine et al. 1985, Paine and Trimble 2004). The authors considered this to be a long-term experiment, but there is only one plot in which seastars were removed and four control plots. Following an El Niño event, mussels, which can out-compete seaweeds for space, dominated the removal plot; macroalgae, which were at first killed off by the warm water, rebounded in the control plots. Paine and Trimble (2004) state that this is evidence for multiple stable states, but their result relies on a press perturbation. They are also assuming that sudden shifts over time are evidence for discontinuous jumps.

Barkai and McQuaid (1988) reported a similar sudden shift in lobsters and associated species on two islands, Marcus Island and Malgas Island, off the west coast of South Africa. Up until the late 1960s, both islands had rock lobsters (*Jasus lalandii*) and seaweeds with low densities of several species of whelk (*Burnupena* spp) and mussels. Rock lobsters eat mussels and whelks, keeping their densities low, and the lack of mussels allowed seaweeds to persist. There was then a crash in rock lobster populations on Marcus Island, and whelks and mussels became common. Malgas Island continued to be dominated by seaweeds and rock lobsters (*Jasus lalandii*). Lobsters transplanted to Marcus Island are quickly overwhelmed and eaten by the whelks, which is very unusual. Barkai and McQuaid suggest the pattern of persistence is consistent with the idea of alternative stable states.

Work in Australia on the Great Barrier Reef provides a good example of how long-term data can be collected, maintained, and used. The Long-Term Monitoring Program, which is overseen by the Australian Institute of Marine Sciences (AIMS), has been surveying 47 reefs since 1993. The data have been used by a number of researchers to look for shifts from coral to algae. While data are said to be available via the AIMS website, it is very difficult to obtain them without help from one of the researchers involved in the sampling. Ninio et al. (2000) examined at changes from 1992 to 1997 on up to 52 reefs. Note there is a difference between Ninio et al. (2000) and the AIMS websites in the number of reefs and the time interval sampled. Ninio et al. found no evidence for shifts. This may be a reflection of persistence in the current state and cannot be used as evidence against the possibility of multiple stable states in coral reefs.

9.4.2 Examples from lakes

There appears to be a general pattern in lakes that involves a switch between a clear-water state and a turbid-water state. Lakes in a clear-water state contain submerged vascular plants that are attached to the bottom; these plants are often referred as submerged macrophytes or SAV (i.e. submerged attached vegetation). In contrast, lakes in a turbid-water state differ from one ecosystem to the next and can include algae or vascular plants that are not attached to the bottom. Algae may include phytoplankton, periphyton (attached algae), epiphytes, benthic mats, or floating mats; vascular plants in the water column are typically floating mats on the surface.

There are some examples based on time-series data from a single lake, but this is not the norm. Lakes are discrete entities, and so most studies involve a number of small lakes sampled once or over a few years. There are, however, several studies done in single lakes and over long time intervals that are worth mentioning because they document shifts from a turbid to clear state. This is much less common than a shift from clear to turbid.

Roelke et al. (2007) used a 34-year record of physical, chemical, and biological data from Lake Kinneret (the Sea of Galilee) to look for sudden shifts and to infer multiple stable states. These data were generated as part of a long-term monitoring program of the lake and of the Kinneret bleak (*Acanthobrama terraesanctae*) fishery. The Kinneret bleak is related to carp and feeds primarily on zooplankton. The fishery collapsed in 1993–94 and had not recovered by 2002, which is the last year for which Roelke et al. report data. Spring blooms of the dinoflagellate (*Peridinium gatunense*) were rare before the collapse and common after the collapse. There were also additional changes in zooplankton, other species of phytoplankton, and water chemistry. Discriminant analysis of chemical and biological data shows distinct differences before and after the collapse. This is not surprising since the purpose of discriminant analysis is to find differences among predefined groups. Roelke et al. suggested that the data, which show abrupt and persistent shifts in dinoflagellate dynamics, provide evidence for multiple stable states.

Temporal shifts from turbid to clear water due to bottom-up processes have also been reported. Both Mihaljevic et al. (2010), studying a flood plain lake in Croatia, and O'Farrell et al. (2011), studying a flood plain lake on the Lower Paraná River in Argentina, showed that flooding caused shift from a turbid to a clear state. The shift in the Croatian lake was linked with flooding from the Danube River in 2006. For decades prior to 2006 the lake had been dominated by phytoplankton and afterwards was converted to a lentic lake with submerged macrophyte vegetation. Similarly, O'Farrell et al. showed that over a 10-year period there was a shift between free-floating plants during flood years and cyanobacterial blooms during periods of extreme low water. As we have seen before, both Mihaljevic et al. (2010) and O'Farrell et al. (2011) linked the abrupt changes in species composition to the idea of thresholds and multiple states.

These studies highlight the two dominant views about the causes for the shift between clear and turbid states (Scheffer et al. 1993, Scheffer and van Nes 2007). One view is that the switch is due to changes in nutrients—and thus a bottom-up process. At low nutrient concentrations, submerged plants dominate and the water column, which is relatively clear of phytoplankton, supports a diversity of zooplankton and fishes. As nutrient levels increase, phytoplankton and/or floating mats blossom. A decline in light due to algae and dissolved oxygen due to secondary production (Morris et al. 2004) leads to a die-off of submerged plants and a loss of zooplankton and fishes. On the other hand, top-down cascades have also been invoked as the cause. The loss of apex predators (i.e. piscivorous fishes) leads to a trophic cascade and ultimately blooms of phytoplankton, other algae, and floating mats. There is ample experimental and observational evidence for both processes and for the interactive effect. These include cover by macrophytes or other species (Bayley et al. 2007, Trochine et al. 2011), light levels (Morris et al. 2004), water levels (Loverde-Oliveira et al. 2009, Mihaljevic et al. 2010, O'Farrell et al. 2011), and even pollutants (Sayer et al. 2006, Stansfield et al. 1989). Scheffer and van Nes (2007) review the various drivers of alternative states in shallow lakes in more detail.

9.4.3 Terrestrial examples

Long-term temporal patterns in terrestrial ecosystems tend to be associated with species or land that are of economic value. We have already considered two such cases in Chapter 1—the decline of the mountain ash forests of Australia and the demise of the passenger pigeon in North America. The other terrestrial systems for which long-term data are commonly available are rangelands, and here we will focus on the temporal patterns in the semi-arid grasslands of the south-western USA.

The rangelands of the American south-west are a mixture of shrubs and grasses. Three of the most common shrubs are creosote bush (*Larrea tridentata*), tarbush (*Flourensia cernua*), and mesquite (several species in the genus *Prosopis*). All three shrubs are native. The underlying grasses and forbs provide forage for range animals and the value as rangeland depends on which shrub dominates, with areas dominated by creosote bush providing the least amount of forage and those dominated by tarbush providing the most (Buffington and Herbel 1965). Buffington and Herbel (1965) provide a detailed summary of the literature and changes in the semi-arid grassland of the Jornada Basin Experimental Range. The Jornada Basin lies in southern New Mexico and in 1982 the Jornada Basin LTER site was established by the US National Science Foundation. Buffington and Herbel (1965) summarize the commentaries of botanists from as early as 1848. Using the General Land Office survey notes from 1858 and range surveys done in 1915, 1928, and 1963, Buffington and Herbel document soil types and changes in plant communities and grazing history.

Bahre and Shelton (1993) give similar information for the rangelands in south-eastern Arizona. The site is just west of the Jornada Basin and so is similar in many ways. Bahre and Shelton provide a description of the long-term changes in vegetation since 1870. They noted that Aldo Leopold, who worked for the US Forest Service in southern Arizona in the early 1900s, wrote that ranchers would talk about how mesquite had taken over the rangelands (Leopold 1924).

Observations of these studies and others have been used to develop state-and-transition models and to make inferences about the plausibility of multiple stable states and thresholds. Bestelmeyer and colleagues have continued to develop state-and-transition models based on the Jornada Basin (Bestelmeyer 2006, Bestelmeyer et al. 2003, 2009). Similarly, Archer (1989) used a simulation model and data from Texas to discuss the possibility of multiple states (see also Figure 7.1 for Archer's conceptualization of multiple states). Mashiri et al. (2008) analyzed the effects of grazing, precipitation, and mesquite on grasses in the Santa Rita Experimental Range in southern Arizona, which is within the area discussed by Bahre and Shelton (1993). Mashiri et al. had data for two time intervals—one with 12 years of data and the other with 34 years. They were unable to detect an effect of grazing and admitted that they could not distinguish between equilibrium and non-equilibrium models (i.e. smooth phase shifts versus discontinuous jumps associated with multiple stable states).

There have also been studies of time-series data in tropical forests. One of the most widely cited is Dubin et al.'s (1990) study of fire and elephants in the Serengeti. Fire and elephants cause a decline and lack of recovery of woodlands, and Dubin et al.'s observational and experimental data from the 1960s and 1980s suggest that the lack of recovery is due to hysteresis. However, the evidence is weak. One of the best studies followed succession in rain forest plots for 12 years after hurricane damage (Vandermeer et al. 2004). Vandermeer et al. examined the multivariate distance between plots at successive time points with the expectation that divergence among plots implies divergence into different basins of attraction. They found some evidence for divergence. Interestingly, they equated divergence with a nonequilibrium model and convergence with an equilibrium model and so made the same mis-step as Mashiri et al. (2008).

In another study, pollen cores were used to look for a bimodal distribution of pollen associated with tropical rain forests in the Amazon Basin (da Silveira Lobo Sternberg 2001). Throughout the Last Glacial Maximum the region saw fluctuations in precipitation. Da Silveira Lobo Sternberg constructed a two-state occupancy model for the tropical forest–savanna system in which precipitation in the dry season affected the probability of occupancy. This gave rise to the typical S-shaped curve of a cusp catastrophe. Samples taken at random points along the pollen cores produced a bimodal distribution. This is consistent with two alternative states but it could also reflect two different environmental regimes.

9.5 Evidence from spatial patterns

The interpretation of spatial patterns as evidence for multiple stable states requires us to make two assumptions. The first assumption is that the environment—that is the set of parameters controlling multiple states—is constant over the area being sampled. The simplest situation would be a collection of distinct and similar places—such as a group of small lakes with similar physical and chemical features. The second assumption is more critical. We must assume that these particular and unchanging environmental conditions can support two or more stable states.

The assumption of similar conditions may be expanded a bit when boundaries between different community types are being examined as evidence of multiple stable states. Often a boundary between two community types reflects a gradient in environmental conditions. The simplest explanation for the boundary is a smooth but abrupt shift from one community to the other as one or more critical parameters change and favor one or the other state. However, if the boundary involves multiple stable states, we must assume there is a set of conditions at the interface of the two community types that could support both states. In other words, there is a set of parameters where there is modality, inaccessibility, and possibly hysteresis. We should see a mosaic of the two community types at this interface. Note that a sharp boundary between two communities could arise from priority effects, and so we do not need to assume there is a gradient. However, in many situations there is a gradient and its impact needs to be taken into account.

The problem is that we run right up against the issues raised by Peterson (1984), which we discussed in Chapter 2. On one hand, if we find no differences in the underlying environmental conditions we cannot rule out that we missed finding a critical parameter that accounts for a smooth shift between states. On the other hand, differences in environmental conditions could arise from the activities of the species involved, for example changes in soil conditions because of leaf litter, and so we cannot assume that differences in environmental conditions rule out the possibility of multiple stable states. This is exactly the reason why Peterson argued that we must show that the very same site can be occupied by different community types.

Given these issues it is unlikely that spatial patterns of species that depend on ecosystem engineers can provide plausible evidence for multiple stable states. Ecosystem engineers are known to alter environmental conditions, and so it would be very difficult to discern causality. Did different species give rise to different environments or did different environments favor different species? There could be multiple stable states in the former case, and definitely not in the latter case. Moreover, researchers are unlikely to go looking for multiple states if environmental conditions differ. The simplest explanation is that different environments favor different communities. So why look any closer? It therefore should not be a surprise that there are very few examples from terrestrial systems and lots of examples from lake systems. We will consider two better known terrestrial examples and then sample some of the examples from lake systems.

9.5.1 Two terrestrial examples

Our first example involves humans and elephants in Zimbabwe (Hoare and du Toit 1999). Hoare and du Toit do not mention alternative or multiple states, but their results are interesting in the light of Dublin et al.'s (1990) paper, which has been widely cited as an example of multiple states. Hoare and du Toit examined census data on elephants and humans in the Sebungwe region, which covers about 15,200 km^2 in north-western Zimbabwe. They found that elephant density was unrelated to human density, but above 15.6 people per km^2, there were no elephants in their census plots. They suggest that elephants migrate out of areas where humans are common and discuss their results in terms of a threshold at 15.6 people per km^2 and a transition between two states.

A close look at Hoare and du Toit's Figure 2 shows a tantalizing suggestion of hysteresis (Figure 9.1). While Hoare and du Toit did not do so, we think it would be more profitable to view human density as a parameter similar to stock density in the familiar grazing models of May (1977) and Noy-Meir (1975, 1981). Elephant density is the state variable and there are two states—high and low density. These are not the states envisioned by Hoare and du Toit. There appears to be a cusp catastrophe with folds at about 3.5 and 13.5–15.6 people per km^2.

Adema et al.'s (2002) study of dune slacks on the coast of the Netherlands shows how difficult it can be to separate environmental conditions from community states. Dune

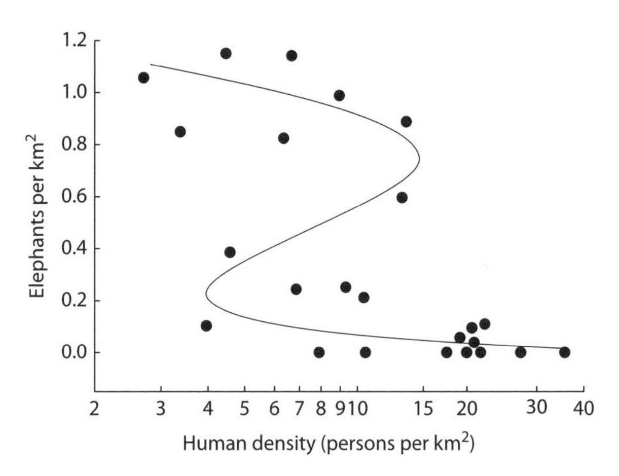

Figure 9.1 Relationship between elephant and human density in the Sebungwe region of Zimbabwe. Folds of the cusp catastrophe were drawn by eye. Data points redrawn from Hoare, R. E. and du Toit, J. T. (1999). Coexistence between people and elephants in African savannas. *Conservation Biology*, 13, 633–639. Reproduced with permission from John Wiley and Sons Ltd.

slacks are low areas between dunes which often contain distinct plant communities, and Adema et al. examined three plant associations found in dune slacks on the Frisian island Texel. They collected data on hydrology, soil characteristics, and vegetation and using detrended correspondence analysis and found clear differences among the community types based on vegetation. They also observed a sharp boundary between two of the communities and suggested that alternative states "occur only if a positive-feedback mechanism is operating." The observational study was followed up by some experimental work (Adema and Grootjans 2003). While the pattern is quite compelling, it is not possible to rule out that the differences among community types are driven by small-scale environmental differences.

9.5.2 Examples from lakes

Patterns in shallow lakes provide some of the best examples of spatial patterns because lakes of this sort are numerous and often many are found under seemingly identical environmental conditions, but with different states. Many studies involve large-scale surveys that combine spatial aspects of using a collection of lakes and temporal aspects by including some information about changes over time. The findings of large-scale surveys tend to be diverse and provide evidence of both top-down and bottom-up control as we have already seen in our earlier mention of Schallenberg and Sorrell's (2009) study of 95 New Zealand lakes in which shifts between clear and turbid conditions were correlated with deforestation and the introduction of exotic fishes.

Studies prior to the mid 2000s usually included fewer than 50 lakes. For example, Jackson (2003) examined data on the physical conditions of turbid and clear states in 30 lakes in Alberta, Canada. He linked the pattern to alternative states, but with little supporting evidence beyond the fact that clear and turbid lakes differed. Jones and Sayer (2003) examined 17 shallow lakes in Norfolk, UK that were dominated by macrophytes but that varied in nutrient inputs and fish densities. Jones and Sayer suggested that control was top-down and due to fishes, and not due to the bottom-up effects of nutrients Interestingly they assume stochastic processes are required for the existence of alternative stable states and assert, "A nutrient-mediated increase in periphyton... is often described as being responsible for the loss of plants from shallow lakes, yet this violates the stochastic assumptions of alternative equilibria." As we have mentioned in previous chapters, the theory of alternative stable states does not require the inclusion of stochastic processes.

By 2007 the number of lakes in the typical survey study doubled. Van Geest et al. (2007) studied 70 lakes in the Netherlands from 1996 to 2004 and found that lakes were either in a macrophyte-dominated or a turbid state. They also observed many switches between the two states. They found no relation with nutrients but found the macrophyte state was associated with low water. They also concluded that the macrophyte state is a long transient and not a true alternative stable state. Bayley et al. (2007) in two separate studies examined the effects of macrophyte cover and total phosphorus on alternative states in shallow lakes in Alberta, Canada. One study used data from 24 lakes taken over 6 years and the other surveyed 82 lakes with data from 1984 to 2003. Some of the lakes in the second study were followed for 20 years. Some 20–29% of the lakes remained in the clear water state and were declared to be stable. The remaining lakes switched at least once, and many switched back and forth multiple times. Switches depended on both phosphorus levels and macrophyte cover; the switch from clear to turbid conditions occurred at higher levels of phosphorus in lakes with more cover. Interestingly, Bayley et al. explicitly linked the switch to a single threshold in phosphorus levels (see their Figure 4), but provide no evidence for two thresholds, which we would expect if there was a cusp catastrophe and hysteresis. Similarly, Sondergaard et al. (2008) suggested that phosphorus loading is important in lack of persistence of a clear water state in their survey of 36 Danish lakes. Sondergaard et al. also reviewed the effects of fish removal; they found that removal increased water clarity but the effect did not persist unless removal continued. Thus removal must be viewed as a press perturbation, suggesting that clear and turbid states lie on a continuum of a single continuous phase shift rather than two unique and stable states. Finally, along the same lines Zimmer et al. (2009) undertook a multivariate analysis of 72 shallow lakes in Minnesota. The lakes clustered into clear and turbid groups and showed some switching over time, but again there was no evidence of modality or the existence of two thresholds that are required for hysteresis.

10

Where do we go from here?

Natural ecosystems have become increasingly fragile as an unintended consequence of human activity, and some systems—as diverse as lakes, coral reefs, grasslands, and fisheries—have experienced dramatic and difficult to reverse shifts in species composition (May 1977, Scheffer et al. 2001, Scheffer and Carpenter 2003). Yet, as we have seen, it is still not clear in many cases if these changes are smooth phase shifts due to undetected changes in environmental conditions or represent true multiple stable states. It remains an open question if multiple stable states are common in nature, but their occurrence implies that past events—both natural and anthropogenic ones—could play a profound role in structuring present-day assemblages.

10.1 What do we know?

The two major lines of research on multiple stable states have been experimental tests and modeling that is linked to patterns in nature, and we have seen that neither method has provided clear-cut answers. Experimental approaches have focused on demonstrating the possibility of two states in the same environment, the role of priority effects, and the existence of hysteresis. Experiments designed to test if communities composed of different species can exist in the same environment and priority effects are based on our modern metaphor of different basins of stability and were first articulated in an exchange of ideas among Connell, Sousa, and Peterson in the 1980s (Connell and Sousa 1983, Peterson 1984, Sousa and Connell 1985). This remains the gold standard of experimental tests of multiple stable states, and it worth restating Peterson's (1984) four criteria. Experimental tests must involve a pulse perturbation, be conducted at a single place, show that two or more different communities can occur in that place, and the communities must be stable in some sense. There are difficulties with meeting all four conditions, but showing a system is stable has been especially daunting (Grimm et al. 1992, Grimm and Wissel 1997).

Tests for different basins of attraction fall into two subclasses. The first consists of the few experiments that directly test for the existence of two or more unique species assemblages in the same environment and that these assemblages can persist for at least one turnover. Connell and Sousa (1983) and more recent reviews (Petraitis and Dudgeon 2004, Schröder et al. 2005) have found very little direct experimental evidence for this sort of test. The second group is larger, but depends on a weaker test. These are experiments which attempt to show that changes in ecological processes, such as recruitment rates of various species after a fire, can lead to the development of different species assemblages. Tests showing priority effects also fall into this class. However, this approach assumes that the initial rates observed after a disturbance will be consistent and continue for long enough to drive the system to a specific basin of attraction. Detection of divergence towards a different basin of attraction naively assumes that the vector defining the initial trajectory of species abundances points towards the equilibrium point. This is often not the case (Fukami and Nakajima 2011, Samuels and Drake 1997). Even so, short-term experiments provide important insights into the existence of multiple stable states.

Hysteresis has also been consistently offered as not only the underlying cause of dramatic and difficult to reverse shifts that have been seen in lakes, coral reefs, grasslands, and fisheries but also as a test for multiple stable states (May 1977, Scheffer et al. 1993, 2001). This is a less inclusive approach because it is possible to have multiple stable states without hysteresis. As we saw in Chapter 2, it is also very difficult to test for hysteresis. While demonstrating that a system has hysteresis is good evidence for multiple stable states, the converse is not necessarily true. The lack of hysteresis does not preclude multiple stable states. We should also remember that the rapidity of a change does not tell us if the shift is a discontinuous jump as expected with hysteresis or a smooth phase shift. The S-shaped curve of a cusp catastrophe is a statement about the position of the equilibrium points and not about the speed at which they are approached. Regime shifts, which may be nothing more that abrupt but smooth changes, cannot be considered prima facie evidence for multiple states.

Modeling is often used when ecosystem-level experiments are not possible or are unethical, or for ecosystems that are economically important and require management guidelines. In the best cases, parameter values for a model are derived from field data and then the model is tested against a second, independent set of data for the likelihood of producing multiple stable states. Modeling has strengths and weaknesses. Models can provide the most compelling examples of plausibility of multiple stable states, but they are often difficult to confirm. Modeling efforts ideally include alternative scenarios that provide clear-cut predictions and explore model robustness against environmental variation in parameters, but as we have seen this is rarely if ever done.

All approaches—from empirical studies of different basins of attraction and hysteresis to models of multiple stable states—tend to be plagued with the difficulties of translating metaphors into clear unambiguous tests. We saw how one concept—the state-and-transition formulation—became inadvertently entangled with the idea of multiple stable states. It is quite clear that the state-and-transition viewpoint was developed as a protocol to manage systems and not as a specific model (Westoby et al. 1989, p. 269). The same it true for the idea that thresholds are a hallmark of multiple stable states, and vice versa. Yet our critique and models show that the existence of multiple stable states does not require threshold responses in parameters. Indeed, multiple stable states can arise in systems in which parameters change in a smooth fashion with changes in environmental conditions. Thus demonstrations that ecosystems show abrupt shifts in parameters, species composition, or other state variables with small changes in environmental conditions are not sufficient tests for multiple stable states. The same holds true for suggestions that ecosystem engineers, positive feedbacks, environmental switches, and stressful abiotic conditions predispose a system to having multiple stable states. These features may be found in systems with multiple stable states but they are not a requirement.

There are some characteristics that are likely to make specific systems good candidates for undertaking experimental studies of multiple stable states. These features include the presence of ecosystem engineers, variation in early successional events, and ecological processes that occur on manageable temporal and spatial scales. Our suggestion that the presence of ecosystem engineers might be a useful may seem to run counter to our comments that their presence does not predispose a system to having multiple stable states. Yet ecosystem engineers can have large and obvious effects on communities and so they are more likely to be good indicators of a particular state than most other species. Their presence may make it easier to identify distinct states. Early successional events provide the variation needed to drive divergence and to set up priority effects. Without this variation, it is unlikely that different states can get started after a perturbation. Finally, experiments require systems in which species and environmental conditions occur on a manageable scale. Multiple stable states may underlie many of the sudden shifts in marine fisheries but the spatial scale of these systems makes experimentation of the sort required by Peterson's criteria impossible. On the other hand, we should not be surprised that Sutherland's observations on fouling plates are seen as such a successful test (Sutherland 1974, 1981). The plates were small and easy to handle, the recruitment of many of the species was highly variable, and certain species showed strong priority effects.

We should remember that hypothesis testing within the context of experimental tests and the estimation of parameters needed for models pose serious statistical and measurement issues. Convincing tests and reliable estimations depend on the length of time and spatial scale over which experiments and observations are done and the number of observations taken. Choosing the right scale and sample size in turn depends on prior knowledge of natural variation. For example, for Peterson's (1984) criteria, how do we define "same," "different," and "self-replacing" if we do not have benchmarks that are extrinsic to our tests and observations, and some understanding of the levels of natural variation? In the remaining two sections we will examine the use of BACI designs to deal with natural variation and argue for more transparency in terminology and greater access to data as ways to move forward.

10.2 Using Before–After–Impact–Control (BACI) designs to test for multiple stable states

The problem of undetected environmental variation in tests for multiple stable states can be addressed through the use of BACI designs (Petraitis and Dudgeon 2004). A BACI design has two time intervals, which are Before and After an impact has been applied or observed and two treatment levels, which are Control and Impact (Figure 10.1). To make use of this design in experiments on multiple stable states we would have an experiment that is initially run for a specified duration and in which a pulse perturbation is applied to some replicates while others are left untouched as controls. We might also include small and large perturbations with the expectation that small perturbations would fail to push the system to an alternative state while large perturbations would be successful. This is the basis of Schröder et al.'s (2005) test of nonrecovery and the idea

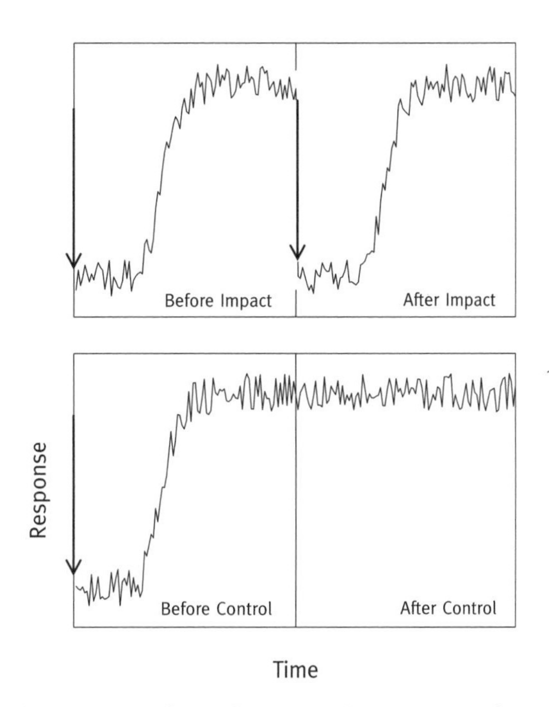

Figure 10.1 Successional trajectories after pulse manipulations in a Before–After–Impact–Control design. The Before panels show succession after initial pulse manipulations were done in all replicates. The After panels show succession after half the replicates received a second perturbation (After–Impact) and half were left untouched (After–Control). The arrows show initial pulse perturbation in all replicates and the second pulse perturbation in the impact replicates. This particular outcome—return to the same state after the second round of perturbation—suggests there is not an alternative state. Figure redrawn from Petraitis, P. S. and Dudgeon, S. R. (2004). Detection of alternative stable states in marine communities. *Journal of Experimental Marine Biology and Ecology* 300, 343–371. Reproduced with permission from Elsevier.

behind the use of perturbations of different sizes in Petraitis's experimental design (Petraitis and Latham 1999, Petraitis et al. 2009). Then a fraction—usually half—of the replicates is reset to initial conditions, which is the After interval. If different communities observed at the end of the Before interval are truly alternative states and not the result of undetected site-specific differences, then resetting experimental plots to initial conditions should produce different outcomes at the end of the After interval. Typically experiments testing multiple stable states only do the first half, that is the Before interval. We know of only one attempt to redo an experiment and include the After interval; Petraitis and Dudgeon rescraped half of their experimental clearings in 2010 for a Before interval of 13 years (Petraitis et al. 2009) and are now in the process of following the ongoing successional changes.

This BACI design includes at least nine possible levels of treatment (Table 10.1). These include replicates that are never subjected to pulse perturbations throughout the entire course of the experiment and thus are always controls (labeled CC in Table 10.1). There are replicates that are perturbed in the same fashion at the start of both intervals (SS and LL), and there are also replicates for which the type of perturbation or lack thereof is not the same in both intervals (CS, CL, SC, SL, LC, and LS). For each treatment level, there are two sets of data. The first are the data collected during the Before interval and the second are the data collected during the After interval.

This design allows us to construct a very straightforward test for multiple stable states and also to determine if the patterns we observe are the result of undetected environmental variation. There are obviously many possible tests given that there are 18 sets of data (nine treatment levels × two time intervals), but we can gain a sense of how we might disentangle undetected site-specific variation from multiple stable states by considering just a few.

Table 10.1 List of possible manipulations and the data collected from a Before–After–Impact–Control (BACI) design to test for multiple states.

Manipulation at the start of the		Data collected during the	
Before interval	**After interval**	**Before interval**	**After interval**
Control	Left as is	$(CC)_B$	$(CC)_A$
Control	Small perturbation	$(CS)_B$	$(CS)_A$
Control	Large perturbation	$(CL)_B$	$(CL)_A$
Small perturbation	Left as is	$(SC)_B$	$(SC)_A$
Small perturbation	Small perturbation	$(SS)_B$	$(SS)_A$
Small perturbation	Large perturbation	$(SL)_B$	$(SL)_A$
Large perturbation	Left as is	$(LC)_B$	$(LC)_A$
Large perturbation	Small perturbation	$(LS)_B$	$(LS)_A$
Large perturbation	Large perturbation	$(LL)_B$	$(LL)_A$

Two of the more interesting comparisons are a test for resilience when subjected to small perturbations and a test of divergence when subjected to large perturbations. Assume that we are examining the trajectory over time or the endpoint of a species whose abundance is a proxy for one of two or more multiple states. If a small perturbation does not tip the system, we would expect the trajectory and endpoint to be similar in the Before and After intervals. That is, the null hypothesis is

$$H_0 : (SS)_B = (SS)_A$$

where the subscripts identify the Before and After data. In contrast, if a large perturbation tips the system to an alternative state, we predict that history would not always repeat itself, and the null hypothesis would be

$$H_0 : (LL)_B = (LL)_A.$$

Rejection of either null hypothesis—and in particular the latter—would suggest that the system contains alternative states and strongly falsifies the hypothesis that undetected site-specific variation is the underlying cause of differences among replicates.

Comparisons among replicates that were switched in their assignment can tell us if the system has "memory" or if there are priority effects. If the Before treatment alters the outcome of the After treatment, then the system has some memory. For example we could test for this among the replicates given the large perturbation during the After interval by examining

$$H_0 : (CL)_A = (SL)_A = (LL)_A$$

For priority effects, we would have a null expectation that order does not matter, for example

$$H_0 : (SL)_A = (LS)_A$$

There are also number of tests that could be done to check for biases. For example, we might wonder if assignment of replicates to treatments in the After interval was without bias. In this case the null hypotheses are

$$\begin{aligned} H_0 &: (CC)_B = (CS)_B = (CL)_B \\ H_0 &: (SC)_B = (SS)_B = (SL)_B \\ H_0 &: (LC)_B = (LS)_B = (LL)_B. \end{aligned}$$

The first equality then implies that species abundance in the replicates from the controls at the end of the Before interval were identical with respect to their assignment to a treatment in the After interval. The second and third equalities are similar statements about replicates in originally assigned to the small and large perturbations, respectively. Along the same lines, we could check for effects of temporal variation by considering

$$H_0 : (CC)_B = (CC)_A$$

Here we would be looking to see if the changes in the controls during the After interval matched the change seen in the Before interval.

These checks for biases, however, present statistical problems because failure to reject does not mean the null is true. Ideally a power analysis should be undertaken prior to the start of the experiment to be confident that there is sufficient power to detect a difference, which is ecologically meaningful, prior to the start of the experiment.

These are but a few of the possible comparisons that may be informative. It is quite likely that comparisons among competing hypotheses will not provide clear-cut, yes or no conclusions, and it may be more productive to use an information-theoretic approach to evaluate the relative support for different hypotheses (Angilletta et al. 2006, Burnham and Anderson 2002).

10.3 Moving towards evidence-based ecology

The small number of good experiments undertaking tests of multiple stable states is a dismal commentary on the state of ecology. Certainly advances in community ecology are slower than in other fields because this sort of ecological research often requires months, or even years, of data collection, but the slower pace means the field is losing ground relative to other areas of biology.

There are two things that could be done to improve the quality and quantity of experimental results. The first is better access to primary data and negative results, and the second is greater transparency about causality so we can make bolder conjectures. What makes this all the more remarkable is that ecologists usually believe that data collection and inferences about causal links are not a problem. Underwood (1990) suggested observations are "easy to make." In another paper, he discussed not only the observation of patterns but also the development of explanations, models, hypotheses, and predictions in less than three pages while spending nearly 27 pages on the development and interpretation of critical tests (Underwood 1991).

The idea of moving scientific discovery forward by making bold conjectures and forming critical tests is old news for ecologists (Underwood 1990, 1991), but we would suggest that the pace forward has been slow because our conjectures have not been so bold. This timidity is not due to a lack of imagination but rather a lack of transparency about causality. In particular, discussions about multiple stable states often involve overly technical jargon, experiments lack clear statements about the falsifiability of hypotheses with no safeguards against confirmation bias, and presentation of unusual results or observations often relies on ad hoc explanations and appeals to long-standing evidence from authorities. It is unsettling that many of these characteristics are well-known indicators of pseudoscience such as intelligent design or astrology (Bunge 1984, 1991, Gruenberger 1964, Langmuir 1989, Lilienfeld and Landfield 2008, Skrabanek 1987).

10.3.1 Access to primary data and negative results

Access to original data is critical for moving any field forward, and the lack of access is one of the leading impediments to progress not only on the question of multiple stable states but also for ecology in general. First and foremost, access to primary data is important because other researchers can use the data to design new experiments, to benchmark their own observations and results, and to improve the quality of comparisons made across various temporal and spatial scales. More often than not, a researcher in the process of designing a new experiment does not have the data in hand that is necessary to make informed decisions on the number and range of treatment levels or on the number of replicates per level (Petraitis 1998). The range of treatment levels is usually chosen in a very arbitrary fashion, but the range should be set by the amount of natural variation. For example, an experiment testing the intraspecific effects of density on the changes in biomass of an organism, the maximum and minimum number of animals per cage or plants per pot should reflect the range of natural densities seen in the field. This sort of information is relatively easy to acquire prior to setting up the experiment, but the next two required pieces—the number of treatment levels and the number of replicates per level—are not so readily available. The number of treatment levels depends on the researcher's hypotheses about how density affects changes in biomass. Is the response expected to be linear, curvilinear, or threshold-like? Detection of a linear response may only need two levels, but a threshold response might require at least five. Finally, the number of replicates per treatment, which in this hypothetical example is the number of cages or pots per treatment level, requires some knowledge of the average variation in changes in biomass in nature. Access to primary data from other studies could often provide enough information to answer these issues and would lead to better designed and more informative experiments.

Access to primary data would also increase the quality and types of comparisons among different studies, and such access would make it possible for researchers to ask if their results and observations are different from previous findings. For experiments on multiple states, summary statistics and metrics such as averages and indices of diversity are routinely reported, but this is often not enough. If someone is interested in changes in variance or skewness, the usual summary statistics would not be sufficient. Finally, new questions and new techniques cannot be anticipated, and the availability of original data would obviate the need for current researchers to second-guess future needs.

Three objections are usually raised about sharing data and providing access. First, we have heard statements to the effect that scientists would never agree to share data. This is not true and reflects a cultural bias on the part of ecologists. In genomics, many journals require posting of nucleotide sequence data in GenBank, an open-access online database, prior to submission of a paper. The submitted paper must give the sequence's accession number, which is assigned by GenBank. Moreover, museums throughout the world have managed and maintained large collections that are freely available. Recently Fisher et al. (2009) reported that the predatory marine snail, *Nucella lapillus*, has increased in size by 22% over the last century. This study would not have been possible without access to extensive museum collections that were made nearly 100 years ago. While ecologists are already taking steps develop databases akin to GenBank, we know of no ecological journal that requires access to the primary data at the time of publication of a paper, as is the norm in genomics.

Second, ecologists often say that questions in ecology are so system or species specific that the sampling methods used in one study are not informative to others, and fields such as genomics does not face this problem. If this were true, it begs the question of whether ecology is anything more than a form of stamp collecting. Moreover, calibration among different methods can be done. Granted, the logistics of sampling under different conditions or in different locations often means the same species is counted in different fashions by different researchers. But if meaningful comparisons cannot be made then how can we draw generalizations?

Finally, the most disturbing claim against allowing access to primary data is that the data belong to the researcher. The collection of ecological data is often supported by funding from national research foundations and ethically belongs not only to the researcher but also the taxpayers who help fund the research. In the USA, the LTER network has a policy that requires researchers to make data "available online with as few restrictions as possible" and with "every effort to release data in a timely fashion and with attention to accurate and complete metadata."

Even the open-access databases now available to ecologists are a step behind what is now the norm in health sciences and social policy. These fields have moved quite strongly towards the approach known as evidence-based practice. The fields of medicine and social policy have started to establish not only norms for evidence but also organizations that undertake systematic reviews of specific treatments and maintain open-access libraries of the results. In the health sciences, it is the Cochrane Collaboration (<http://www.cochrane.org>) and in social policy, it is the Campbell Collaboration (<http://www.campbellcollaboration.org>). Recently it has been suggested that an online repository not only of data and results but also of hypotheses and methods be created; this would be submitted prior to doing the experiment (Schooler 2011). Schooler made this suggestion as a way to answer questions about the decline effect, which is the well-known phenomenon of declining effect sizes as experiments are repeated. Ecology has a long way to go to catch up, even though we could easily argue that the environmental issues of climate change and abrupt irreversible shifts found in systems with multiple stable states have social costs as large if not larger than those associated with assessing the best practices in medicine and social policy.

10.3.2 Making more precise statements about causality, and bolder conjectures

As Darwin hinted at in his letter to Henry Fawcett, mentioned at the beginning of Chapter 9, any geologist, and for us any ecologist, worth his or her salt has an idea in mind at the start of collecting data or setting up an experiment. All researchers have at least a vague notion of causality and the relationship between condition X and outcome Y at the onset of data collection. At some point, a researcher makes conjectures or assumptions about the relationship between X and Y and constructs one or more black boxes between X and Y which are then stated as hypotheses. The black box represents the researcher's conjectures of how X and Y are linked by physical, chemical, and biological processes, and the hypotheses are statements about how the researcher is going to test the predictions that flow from the black box.

Conjectures are based on ancillary information and commonsense, even though, as Einstein warned, "Common sense is the collection of prejudices acquired by the age of eighteen." Commonsense can fail us in the most ordinary situations. If we see a man in a wheelchair on one side of busy divided highway, and an hour later we see him on the other side, commonsense and use of Occam's razor tells us that the man crossed using the crosswalk at the corner. We don't think he rolled out into traffic, threw his chair over the divider and rolled over to the other side. It is too implausible even though it may be possible. But failures in commonsense occur all the time in everyday life. As ecologists, we are deluding ourselves if we do not think that these kinds of failures are common when we construct our black boxes for the complex and variable systems that we find in nature. Moreover, there are serious risks if we do nothing to protect ourselves against such mistakes.

The most productive way to protect ourselves against mistakes and to move forward would be to focus on prying open the black boxes (Skrabanek 1994). Even better, we could set up opposing black boxes based on alternative conjectures that give rise to distinctly different predictions. More precise hypotheses would mean that we are able to make better, or perhaps even bolder, conjectures about what is going on inside the black box.

Making more precise hypotheses requires precision in our use of terms, and we suggest that the study of multiple stable states in ecology would benefit from the use of clearer and more consistent terminology. While the use of some terms, such as stability, will always present problems, it would be a great improvement if terms were placed in a meaningful ecological context (see comments on this in Grimm and Wissel 1997). It would also help if there was more consistency. The study of multiple stable states in ecology is certainly not helped when phrases such as phase shifts, regime shifts, and thresholds are used almost interchangeably and applied to both smooth continuous changes and discontinuous jumps.

One way to improve transparency in terminology is to ask the question, "what are the units?" High-profile papers on multiple stable states abound with words like resilience, and phrases such as slow and fast drivers. What are the units of measurement for resilience, and what makes one driver slow and another one fast? Pinning down the units of measurement has an added benefit in that it will make using open-access datasets easier to use. It is far easier to make comparisons if we know the units of measurement. It is also a great way to weed out junk science.

Consistent and transparent terms that are grounded in easily measurable units provide a way to anchor our predictions using ambient conditions or competing risks as benchmarks. For example, the incidence rate of lung cancer in the USA is about 0.06% per annum, but this is a meaningless rate without some context. Context can be provided by the risk relative to other groups (e.g. US smokers are 10 to 20 times more likely to develop lung cancer than nonsmokers) or relative to other causes (e.g. the number of deaths in the USA per day from lung cancer is 464, equivalent to the number of fatalities from the crash of a 747–400ER jumbo jet per day). Marine ecologists, for example, have been doing experimental studies for at least 50 years, if we take Connell's (1961) work as a starting point, but how often do we hear statements of the sort that mortality due to a competitor is twice as likely as death due to a predator?

Finally, probably the boldest conjecture each of us can make involves asking ourselves the question, "What evidence would it take for me to give up my most cherished hypothesis?" How this question is answered is the core of what separates ecology and all legitimate scientific endeavor from junk science such as intelligent design. There are a couple important wrinkles to this. Firstly, suggesting that the null hypothesis is the way to deal with this question does not wash. Proving a null hypothesis is impossible. In some cases, it is possible to state two alternative hypotheses in such a fashion that in rejecting one of the alternatives we provisionally accept the alternative that there is "no effect." This is routinely done in product safety testing and has been used by ecologists to test for bioequivalence (e.g. Castilla et al. 2004, Mapstone 1995). Proof of safety, however, requires a clear statement about the suspected effect size, which in turn requires a good estimate of the background or natural variation in the effect itself. This is not possible for most ecological studies, and again brings us back to the need for access to primary data. Even when two (or more) clear alternative hypotheses are stated explicitly, the data often do not unequivocally reject all but one of the hypotheses. We would point out that there are ways to deal with this problem head-on rather than avoid it (e.g. Burnham and Anderson 2002). Many researchers shy away from making bold conjectures about how they would go about rejecting their pet ideas because they know that it is a difficult task. However, rather than giving up, we need to make an honest effort at asking what it would take to reject a cherished view of the world.

This is an important question given that we have witnessed large-scale changes and collapses in many ecosystems over the last several decades, and in most cases it is not clear if these dramatic changes are reversible. Moreover, the existence of multiple stable states means that history can play a profound role in structuring present-day assemblages (Knowlton 1992, Lewontin 1969, Petraitis and Dudgeon 1999, Ricklefs 1987). In these ecosystems, history matters and reflects a balance between long-term, large-scale processes such as natural disasters and more local and rapid ecological interactions. We need to understand that balance if we are to make progress in understanding multiple states.

References

Åberg, P. (1992a). A demographic study of two populations of the seaweed *Ascophyllum nodosum*. *Ecology*, 73, 1473–87.

Åberg, P. (1992b). Size based demography of the seaweed *Ascophyllum nodosum* in stochastic environments. *Ecology*, 73, 1488–501.

Acácio, V., Holmgren, M., Rego, F., Moreira, F. and Mohren, G. M. J. (2009). Are drought and wildfires turning Mediterranean cork oak forests into persistent shrublands? *Agroforestry Systems*, 76, 389–400.

Acevedo, M. F. (1981). On Horn's Markovian model of forest dynamics with particular reference to tropical forests. *Theoretical Population Biology*, 19, 230–50.

Adema, E. B. and Grootjans, A. P. (2003). Possible positive-feedback mechanisms: plants change abiotic soil parameters in wet calcareous dune slacks. *Plant Ecology*, 167, 141–9.

Adema, E. B., Grootjans, A. P., Petersen, J. and Grijpstra, J. (2002). Alternative stable states in a wet calcareous dune slack in the Netherlands. *Journal of Vegetation Science*, 13, 107–14.

Allee, W. C. (1931). *Animal aggregations. A study in general sociology.* University of Chicago Press, Chicago.

Allen-Diaz, B. and Bartolome, J. W. (1998). Sagebrush–grass vegetation dynamics: comparing classical and state-transition models. *Ecological Applications*, 8, 795–804.

Andersen, T., Elser, J. J. and Hessen, D. O. (2004). Stoichiometry and population dynamics. *Ecology Letters*, 7, 884–900.

Andersen, T., Carstensen, J., Hernandez-Garcia, E. and Duarte, C. M. (2009). Ecological thresholds and regime shifts: approaches to identification. *Trends in Ecology and Evolution*, 24, 49–57.

Angilletta, M. J., Oufiero, C. E. and Leaché, A. D. (2006). Direct and indirect effects of environmental temperature on the evolution of reproductive strategies: an information-theoretic approach. *The American Naturalist*, 168, E123–E135.

Archer, S. (1989). Have southern Texas savannas been converted to woodlands in recent history? *The American Naturalist*, 134, 545–61.

Armstrong, R. A. (1979). Prey species replacement along a gradient of nutrient enrichment: a graphical approach. *Ecology*, 60, 76–84.

Armstrong, R. A. and McGehee, R. (1980). Competitive exclusion. *The American Naturalist*, 115, 151–70.

Arnol'd, V. I. (1992). *Catastrophe theory.* Springer-Verlag, New York.

Augustine, D. J., Frelich, L. E. and Jordan, P. A. (1998). Evidence for two alternate stable states in an ungulate crazing system. *Ecological Applications*, 8, 1260–9.

Bahre, C. J. and Shelton, M. L. (1993). Historic vegetation change, mesquite increases, and climate in southeastern Arizona. *Journal of Biogeography*, 20, 489–504.

Bakker, E. S., Van Donk, E., Declerck, S. A. J., Helmsing, N. R., Hidding, B. and Nolet, B. A. (2010). Effect of macrophyte community composition and nutrient enrichment on plant biomass and algal blooms. *Basic and Applied Ecology*, 11, 432–9.

Barkai, A. and McQuaid, C. (1988). Predator–prey role reversal in a marine benthic ecosystem. *Science*, 242, 62–4.

Barker, T., Hatton, K., O'Connor, M., Connor, L., Bagnell, L. and Moss, B. (2008). Control of ecosystem state in a shallow, brackish lake: implications for the conservation of stonewort communities. *Aquatic Conservation–Marine and Freshwater Ecosystems*, 18, 221–40.

Bayley, S. E., Creed, I. F., Sass, G. Z. and Wong, A. S. (2007). Frequent regime shifts in trophic states in shallow lakes on the boreal plain: alternative "unstable" states? *Limnology and Oceanography*, 52, 2002–12.

Beisner, B. E., Haydon, D. T. and Cuddington, K. (2003). Alternative stable states in ecology. *Frontiers in Ecology and the Environment*, 1, 376–82.

Bender, E. A., Case, T. J. and Gilpin, M. E. (1984). Perturbation experiments in community ecology—theory and practice. *Ecology*, 65, 1–13.

Bertness, M. D., Trussell, G. C., Ewanchuk, P. J. and Silliman, B. R. (2002). Do alternate stable community states exist in the Gulf of Maine rocky intertidal zone? *Ecology*, 83, 3434–48.

Bertness, M. D., Trussell, G. C., Ewanchuk, P. J. and Silliman, B. R. (2004a). Do alternate stable community states exist in the Gulf of Maine rocky intertidal zone? Reply. *Ecology*, 85, 1165–7.

Bertness, M. D., Trussell, G. C., Ewanchuk, P. J., Silliman, B. R. and Crain, C. M. (2004b). Consumer-controlled community states on Gulf of Maine rocky shores. *Ecology*, 85, 1321–31.

Bestelmeyer, B. T. (2006). Threshold concepts and their use in rangeland management and restoration: the good, the bad, and the insidious. *Restoration Ecology*, 14, 325–9.

Bestelmeyer, B. T., Brown, J. R., Havstad, K. M., Alexander, R., Chavez, G. and Herrick, J. (2003). Development and use of state-and-transition models for rangelands. *Journal of Range Management*, 56, 114–26.

Bestelmeyer, B. T., Tugel, A. J., Peacock, G. L., Robinett, D. G., Sbaver, P. L., Brown, J. R., Herrick, J. E., Sanchez, H. and Havstad, K. M. (2009). State-and-transition models for heterogeneous landscapes: a strategy for development and application. *Rangeland Ecology and Management*, 62, 1–15.

Boukal, D. S. and Berec, L. (2002). Single-species models of the Allee effect: extinction boundaries, sex ratios and mate encounters. *Journal of Theoretical Biology* 218, 375–94.

Breman, H., Cisse, A. M., Djiteye, M. A. and Elberse, W. T. (1980). Pasture dynamics and forage availability in the Sahel. *Israel Journal of Botany*, 28, 227–51.

Briske, D. D., Fuhlendorf, S. D. and Smeins, F. E. (2003). Vegetation dynamics on rangelands: a critique of the current paradigms. *Journal of Applied Ecology*, 40, 601–14.

Briske, D. D., Fuhlendorf, S. D. and Smeins, F. E. (2006). A unified framework for assessment and application of ecological thresholds. *Rangeland Ecology and Management*, 59, 225–36.

Briske, D. D., Bestelmeyer, B. T., Stringham, T. K. and Shaver, P. L. (2008). Recommendations for development of resilience-based state-and-transition models. *Rangeland Ecology and Management*, 61, 359–67.

Briske, D. D., Washington-Allen, R. A., Johnson, C. R., Lockwood, J. A., Lockwood, D. R., Stringham, T. K. and Shugart, H. H. (2010). Catastrophic thresholds: a synthesis of concepts, perspectives, and applications. *Ecology and Society*, 15(3) [online] <http://www.ecologyandsociety.org/vol15/iss3/art37>

Buchsbaum, R., Valiela, I., Swain, T., Dzierzeski, M. and Allen, S. (1991). Available and refractory nitrogen in detritus of coastal vascular plants and macroalgae. *Marine Ecology–Progress Series*, 72, 131–43.

Buffington, L. C. and Herbel, C. H. (1965). Vegetational changes on a semidesert grassland range from 1858 to 1963. *Ecological Monographs*, 35, 139–64.

Bunge, M. (1984). What is pseudoscience? *The Skeptical Inquirer*, 9, 36–46.

Bunge, M. (1991). What is science—does it matter to distinguish it from pseudoscience—reply. *New Ideas in Psychology*, 9, 245–83.

Burnham, K. P. and Anderson, D. R. (2002). *Model selection and multi-model inference. A practical information-theoretic approach.* Springer-Verlag, New York.

Carpenter, S. R. and Brock, W. A. (2006). Rising variance: a leading indicator of ecological transition. *Ecology Letters*, 9, 311–18.

Carpenter, S. R., Brock, W. A., Cole, J. J., Kitchell, J. F. and Pace, M. L. (2008). Leading indicators of trophic cascades. *Ecology Letters*, 11, 128–38.

Casti, J. (1982). Catastrophes, control and the inevitability of spruce budworm outbreaks. *Ecological Modelling*, 14, 293–300.

Castilla, J. C., Lagos, N. A. and Cerda, M. (2004). Marine ecosystem engineering by the alien ascidian *Pyura praeputialis* on a mid-intertidal rocky shore. *Marine Ecology–Progress Series*, 268, 119–30.

Chase, J. M. (1999). Food web effects of prey size refugia: variable interactions and alternative stable equilibria. *The American Naturalist*, 154, 559–70.

Chase, J. M. (2003a). Experimental evidence for alternative stable equilibria in a benthic pond food web. *Ecology Letters*, 6, 733–41.

Chase, J. M. (2003b). Community assembly: when should history matter? *Oecologia*, 136, 489–98.

Chase, J. M. (2003c). Strong and weak trophic cascades along a productivity gradient. *Oikos*, 101, 187–95.

Chesson, P. L. and Ellner, S. (1989). Invasibility and stochastic boundedness in monotonic competition models. *Journal of Mathematical Biology*, 27, 117–38.

Chesson, P. L. and Warner, R. R. (1981). Environmental variability promotes coexistence in lottery competitive systems. *The American Naturalist*, 117, 923–43.

Chipperfield, P. N. J. (1953). Observations on the breeding and recruitment of *Mytilus edulis* (L.) in British waters. *Journal of the Marine Biological Association of the United Kingdom*, 32, 449–76.

Chisholm, R. A. and Filotas, E. (2009). Critical slowing down as an indicator of transitions in two-species models. *Journal of Theoretical Biology*, 257, 142–9.

Cingolani, A. M., Noy-Meir, I. and Diaz, S. (2005). Grazing effects on rangeland diversity: a synthesis of contemporary models. *Ecological Applications*, 15, 757–73.

Clark, C. W. (1976). *Mathematical bioeconomics: the optimal arrangement of renewable resources.* John Wiley, New York.

Clements, F. E. (1916). *Plant succession: an analysis of the development of vegetation.* Carnegie Institution of Washington Publication no. 242. Carnegie Institution, Washington, DC.

Cole, B. J. (1983). Assembly of mangrove ant communities—patterns of geographical-distribution. *Journal of Animal Ecology*, 52, 339–47.

Cole, R. and McBride, G. (2004). Assessing impacts of dredge spoil disposal using equivalence tests: implications of a precautionary (proof of safety) approach. *Marine Ecology–Progress Series*, 279, 63–72.

Colgan, P. W., Nowell, W. A. and Stokes, N. W. (1981). Spatial-aspects of nest defense by pumpkinseed sunfish (*Lepomis gibbosus*)—stimulus features and an application of catastrophe-theory. *Animal Behaviour*, 29, 433–42.

Collie, J. S., Richardson, K. and Steele, J. H. (2004). Regime shifts: can ecological theory illuminate the mechanisms? *Progress in Oceanography*, 60, 281–302.

Connell, J. H. (1961). Influence of interspecific competition and other factors on distribution of barnacle *Chthamalus stellatus*. *Ecology*, 42, 710–23.

Connell, J. H. (1983). On the prevalence and relative importance of interspecific competition—evidence from field experiments. *The American Naturalist*, 122, 661–96.

Connell, J. H. and Sousa, W. P. (1983). On the evidence needed to judge ecological stability or persistence. *The American Naturalist*, 121, 789–824.

Contamin, R. and Ellison, A. M. (2009). Indicators of regime shifts in ecological systems: what do we need to know and when do we need to know it? *Ecological Applications*, 19, 799–816.

Csada, R. D., James, P. C. and Espie, R. H. M. (1996). The "file drawer problem" of non-significant results: does it apply to biological research? *Oikos*, 76, 591–3.

D'Odorico, P., Laio, F. and Ridolfi, L. (2006). A probabilistic analysis of fire-induced tree-grass coexistence in savannas. *The American Naturalist*, 167, E79–E87.

Dakos, V., Scheffer, M., van Nes, E. H., Brovkin, V., Petoukhov, V. and Held, H. (2008). Slowing down as an early warning signal for abrupt climate change. *Proceedings of the National Academy of Sciences of the United States of America*, 105, 14308–12.

Dakos, V., van Nes, E. H., Donangelo, R., Fort, H. and Scheffer, M. (2010). Spatial correlation as a leading indicator of catastrophic shifts. *Theoretical Ecology*, 3, 163–74.

Daley, C. (1960). *The story of Gippsland.* Whitcombe & Tombs, Melbourne.

Dallman, P. R. (1998). *Plant life in the world's Mediterranean climates.* University of California Press, Berkeley, CA.

Davenport, D. W., Breshears, D. D., Wilcox, B. P. and Allen, C. D. (1998). Viewpoint: sustainability of piñon-juniper ecosystems—a unifying perspective of soil erosion thresholds. *Journal of Range Management*, 51, 231–40.

Deakin, M. A. B. (1990). Catastrophe modeling in the biological sciences. *Acta Biotheoretica*, 38, 3–22.

DeAngelis, D. L. and Waterhouse, J. C. (1987). Equilibrium and nonequilibrium concepts in ecological models. *Ecological Monographs*, 57, 1–21.

Dent, C. L., Cumming, G. S. and Carpenter, S. R. (2002). Multiple states in river and lake ecosystems. *Philosophical Transactions of the Royal Society B: Biological Sciences*, 357, 635–45.

Didham, R. K., Watts, C. H. and Norton, D. A. (2005). Are systems with strong underlying abiotic regimes more likely to exhibit alternative stable states? *Oikos*, 110, 409–16.

Diehl, S. (2007). Paradoxes of enrichment: effects of increased light versus nutrient supply on pelagic producer–grazer systems. *The American Naturalist*, 169, E173–E191.

Dijkstra, H. A. and Weijer, W. (2005). Stability of the global ocean circulation: basic bifurcation diagrams. *Journal of Physical Oceanography*, 35, 933–48.

Ditlevsen, P. D. and Johnsen, S. J. (2010). Tipping points: early warning and wishful thinking. *Geophysical Research Letters*, 37, L19703.

Dixon, P. M. and Pechmann, J. H. K. (2005). A statistical test to show negligible trend. *Ecology*, 86, 1751–6.

Doak, D. F., Estes, J. A., Halpern, B. S., Jacob, U., Lindberg, D. R., Lovvorn, J., Monson, D. H., Tinker, M. T., Williams, T. M., Wootton, J. T., Carroll, I., Emmerson, M., Micheli, F. and Novak, M. (2008). Understanding and predicting ecological dynamics: are major surprises inevitable? *Ecology*, 89, 952–61.

Donangelo, R., Fort, H., Dakos, V., Scheffer, M. and Van Nes, E. H. (2010). Early warnings for catastrophic shifts in ecosystems: comparison between spatial and temporal indicators. *International Journal of Bifurcation and Chaos*, 20, 315–21.

Done, T. J. (1992). Phase-shifts in coral-reef communities and their ecological significance. *Hydrobiologia*, 247, 121–32.

Drake, J. A. (1990). The mechanics of community assembly and succession. *Journal of Theoretical Biology*, 147, 213–33.

Drake, J. A. (1991). Community-assembly mechanics and the structure of an experimental species ensemble. *The American Naturalist*, 137, 1–26.

Drake, J. A., Flum, T. E., Witteman, G. J., Voskuil, T., Hoylman, A. M., Creson, C., Kenny, D. A., Huxel, G. R., Larue, C. S. and Duncan, J. R. (1993). The construction and assembly of an ecological landscape. *Journal of Animal Ecology*, 62, 117–30.

Drake, J. A., Flum, T. and Huxel, G. R. (1994). On defining assembly space—a reply. *Journal of Animal Ecology*, 63, 488–9.

Drake, J. A., Huxel, G. R. and Hewitt, C. L. (1996). Microcosms as models for generating and testing community theory. *Ecology*, 77, 670–7.

Drake, J. M. and Griffen, B. D. (2010). Early warning signals of extinction in deteriorating environments. *Nature*, 467, 456–9.

Droop, M. R. (1974). Nutrient status of algal cells in continuous culture. *Journal of the Marine Biological Association of the United Kingdom*, 54, 825–55.

Dublin, H. T., Sinclair, A. R. E. and McGlade, J. (1990). Elephants and fire as causes of multiple stable states in the Serengeti mara woodlands. *Journal of Animal Ecology*, 59, 1147–64.

Duckstein, L., Casti, J. and Kempf, J. (1979). Modeling phytoplankton dynamics using catastrophe theory. *Water Resources Research*, 15, 1189–94.

Dudgeon, S. and Petraitis, P. S. (2001). Scale-dependent recruitment and divergence of intertidal communities. *Ecology*, 82, 991–1006.

Dunham, A. E. and Beaupre, S. J. (1998). Ecological experiments: scale, phenomenology, mechanism, and the illusion of generality. In W. J. Resetarits Jr and J. Bernardo (eds) *Experimental ecology—issues and perspectives*, pp. 27–49. Oxford University Press, Oxford.

Dyksterhuis, E. J. (1949). Ecological principles in range evaluation. *Botanical Review*, 24, 253–72.

Edwards, A. W. F. (1976). *Likelihood: an account of the statistical concept of likelihood and its application to scientific inference.* Cambridge University Press, Cambridge, UK.

Ellis, J. E. and Swift, D. M. (1988). Stability of African pastoral ecosystems—alternate paradigms and implications for development. *Journal of Range Management*, 41, 450–9.

Englund, G., Sarnelle, O. and Cooper, S. D. (1999). The importance of data-selection criteria: meta-analyses of stream predation experiments. *Ecology*, 80, 1132–41.

Eppinga, M. B., Rietkerk, M., Wassen, M. and De Ruiter, P. C. (2009). Linking habitat modification to catastrophic shifts and vegetation patterns in bogs. *Plant Ecology*, 200, 53–68.

Estberg, G. N. and Patten, B. C. (1976). The relation between sensitivity and persistence under small perturbations. In S. A. Levin (ed.) *Ecosystem analysis and prediction*, pp. 151–4. Society for Industrial and Applied Mathematics, Philadelphia.

Fairweather, P. G. and Underwood, A. J. (1983). The apparent diet of predators and biases due to different handling times of their prey. *Oecologia*, 56, 169–79.

Feng, J. F., Wang, H. L., Huang, D. W. and Li, S. P. (2006). Alternative attractors in marine ecosystems: a comparative analysis of fishing effects. *Ecological Modelling*, 195, 377–84.

Fernandez-Gimenez, M. E. and Allen-Diaz, B. (1999). Testing a non-equilibrium model of rangeland vegetation dynamics in Mongolia. *Journal of Applied Ecology*, 36, 871–85.

Feudel, U. (2008). Complex dynamics in multistable systems. *International Journal of Bifurcation and Chaos*, 18, 1607–26.

Fisher, J. A. D., Rhile, E. C., Liu, H. and Petraitis, P. S. (2009). An intertidal snail shows a dramatic size increase over the past century. *Proceedings of the National Academy of Sciences of the United States of America*, 106, 5209–12.

Fogarty, M. J. and Murawski, S. A. (1998). Large-scale disturbance and the structure of marine system: fishery impacts on Georges Bank. *Ecological Applications*, 8, S6–S22.

Folke, C., Carpenter, S., Walker, B., Scheffer, M., Elmqvist, T., Gunderson, L. and Holling, C. S. (2004). Regime shifts, resilience, and biodiversity in ecosystem management. *Annual Review of Ecology Evolution and Systematics*, 35, 557–81.

Foster, M. S., Nigg, E. W., Kiguchi, L. M., Hardin, D. D. and Pearse, J. S. (2003). Temporal variation and succession in an algal-dominated high intertidal assemblage. *Journal of Experimental Marine Biology and Ecology*, 289, 15–39.

Francis, G. (2012). Too good to be true: publication bias in two prominent studies from experimental psychology. *Psychonomic Bulletin and Review*, 19, 151–6.

Frank, P. W. (1968). Life histories and community stability. *Ecology*, 49, 355–7.

Friedel, M. H. (1991). Range condition assessment and the concept of thresholds—a viewpoint. *Journal of Range Management*, 44, 422–6.

Fukami, T. (2001). Sequence effects of disturbance on community structure. *Oikos*, 92, 215–24.

Fukami, T. and Nakajima, M. (2011). Community assembly: alternative stable states or alternative transient states? *Ecology Letters*, 14, 973–84.

Fung, T., Seymour, R. M. and Johnson, C. R. (2011). Alternative stable states and phase shifts in coral reefs under anthropogenic stress. *Ecology*, 92, 967–82.

Gaffney, P. M. (1975). Roots of the niche concept. *The American Naturalist*, 109, 490.

Gause, G.F. (1934). *The struggle for existence.* Williams and Wilkins Co., Baltimore.

van Geest, G. J., Coops, H., Scheffer, M. and van Nes, E. H. (2007). Long transients near the ghost of a stable state in eutrophic shallow lakes with fluctuating water levels. *Ecosystems*, 10, 36–46.

George, M. R., Brown, J. R. and Clawson, W. J. (1992). Application of nonequilibrium ecology to management of Mediterranean grasslands. *Journal of Range Management*, 45, 436–40.

Gilmore, R. (1981). *Catastrophe theory for scientists and engineers.* John Wiley and Sons, New York.

Gilpin, M. E. and Ayala, F. J. (1973). Global models of growth and competition. *Proceedings of the National Academy of Sciences of the United States of America*, 70, 3590–3.

Gould, S. J. (1992). Dinosaurs in the haystack. *Natural History*, 101, 2–13.

Grebogi, C., Ott, E. and Yorke, J. A. (1987). Chaos, strange attractors, and fractal basin boundaries in nonlinear dynamics. *Science*, 238, 632–8.

Grimm, V. and Wissel, C. (1997). Babel, or the ecological stability discussions: an inventory and analysis of terminology and a guide for avoiding confusion. *Oecologia*, 109, 323–34.

Grimm, V., Schmidt, E. and Wissel, C. (1992). On the application of stability concepts in ecology. *Ecological Modelling*, 63, 143–61.

Groffman, P., Baron, J., Blett, T., Gold, A., Goodman, I., Gunderson, L., Levinson, B., Palmer, M., Paerl, H., Peterson, G., Poff, N., Rejeski, D., Reynolds, J., Turner, M., Weathers, K. and Wiens, J. (2006). Ecological thresholds: the key to successful environmental management or an important concept with no practical application? *Ecosystems*, 9, 1–13.

Gruenberger, F. J. (1964). Measure for crackpots. *Science*, 145, 1413–15.

Gurevitch, J. and Hedges, L. V. (1999). Statistical issues in ecological meta-analyses. *Ecology*, 80, 1142–9.

Guttal, V. and Jayaprakash, C. (2007). Impact of noise on bistable ecological systems. *Ecological Modelling*, 201, 420–8.

Guttal, V. and Jayaprakash, C. (2008). Changing skewness: an early warning signal of regime shifts in ecosystems. *Ecology Letters*, 11, 450–60.

Guttal, V. and Jayaprakash, C. (2009). Spatial variance and spatial skewness: leading indicators of regime shifts in spatial ecological systems. *Theoretical Ecology*, 2, 3–12.

Hall, S. R. (2004). Stoichiometrically explicit competition between grazers: species replacement, coexistence, and priority effects along resource supply gradients. *The American Naturalist*, 164, 157–72.

Halliday, T. R. (1980). Extinction of the passenger pigeon *Ectopistes migratorius* and its relevance to contemporary conservation. *Biological Conservation*, 17, 157–62.

Handa, I. T., Harmsen, R. and Jefferies, R. L. (2002). Patterns of vegetation change and the recovery potential of degraded areas in a coastal marsh system of the Hudson Bay lowlands. *Journal of Ecology*, 90, 86–99.

Hardin, G. (1960). The competitive exclusion principle. *Science*, 131, 1292–7.

Hare, S. R. and Mantua, N. J. (2000). Empirical evidence for north Pacific regime shifts in 1977 and 1989. *Progress in Oceanography*, 47, 103–45.

Hassell, M. P. (1978). *The dynamics of arthropod predator–prey systems*. Princeton University Press, Princeton, NJ.

Hastings, A. (2004). Transients: the key to long-term ecological understanding? *Trends in Ecology and Evolution*, 19, 39–45.

Hastings, A. and Wysham, D. B. (2010). Regime shifts in ecological systems can occur with no warning. *Ecology Letters*, 13, 464–72.

Hastings, A., Byers, J. E., Crooks, J. A., Cuddington, K., Jones, C. G., Lambrinos, J. G., Talley, T. S. and Wilson, W. G. (2007). Ecosystem engineering in space and time. *Ecology Letters*, 10, 153–64.

van der Heide, T., van Nes, E. H., Geerling, G. W., Smolders, A. J. P., Bouma, T. J. and van Katwijk, M. M. (2007). Positive feedbacks in seagrass ecosystems: implications for success in conservation and restoration. *Ecosystems*, 10, 1311–22.

Hesseln, H., Rideout, D. B. and Omi, P. N. (1998). Using catastrophe theory to model wildfire behavior and control. *Canadian Journal of Forest Research–Revue Canadienne De Recherche Forestiere*, 28, 852–62.

Hoare, R. E. and du Toit, J. T. (1999). Coexistence between people and elephants in African savannas. *Conservation Biology*, 13, 633–9.

Holling, C. S. (1973). Resilience and stability of ecological systems. *Annual Review of Ecology and Systematics*, 4, 1–23.

Holt, R. D., Grover, J. and Tilman, D. (1994). Simple rules for interspecific dominance in systems with exploitative and apparent competition. *The American Naturalist*, 144, 741–71.

Horan, R. D., Fenichel, E. P., Drury, K. L. S. and Lodge, D. M. (2011). Managing ecological thresholds in coupled environmental–human systems. *Proceedings of the National Academy of Sciences of the United States of America*, 108, 7333–8.

Horn, H. S. and MacArthur, R. H. (1972). Competition among fugitive species in a harlequin environment. *Ecology*, 53, 749–52.

Hotes, S., Grootjans, A. P., Takahashi, H., Ekschmitt, K. and Poschlod, P. (2010). Resilience and alternative equilibria in a mire plant community after experimental disturbance by volcanic ash. *Oikos*, 119, 952–63.

Hsieh, C. H., Anderson, C. and Sugihara, G. (2008). Extending nonlinear analysis to short ecological time series. *The American Naturalist*, 171, 71–80.

Hubbell, S. P. (2001). *The unified neutral theory of biodiversity and biogeography.* Princeton University Press, Princeton, NJ.

Huggett, A. J. (2005). The concept and utility of "ecological thresholds" in biodiversity conservation. *Biological Conservation*, 124, 301–10.

Huisman, J. and Weissing, F. J. (2001a). Biological conditions for oscillations and chaos generated by multispecies competition. *Ecology*, 82, 2682–95.

Huisman, J. and Weissing, F. J. (2001b). Fundamental unpredictability in multispecies competition. *The American Naturalist*, 157, 488–94.

Hurlbert, S. H. (1984). Pseudoreplication and the design of ecological field experiments. *Ecological Monographs*, 54, 187–211.

Hurtt, G. C. and Pacala, S. W. (1995). The consequences of recruitment limitation—reconciling chance, history and competitive differences between plants. *Journal of Theoretical Biology*, 176, 1–12.

Hutchinson, G. E. (1961). The paradox of the plankton. *The American Naturalist*, 95, 137–46.

Illius, A. W. and O'Connor, T. G. (1999). On the relevance of nonequilibrium concepts to arid and semiarid grazing systems. *Ecological Applications*, 9, 798–813.

Irfanullah, H. M. and Moss, B. (2004). Factors influencing the return of submerged plants to a clear-water, shallow temperate lake. *Aquatic Botany*, 80, 177–91.

Irvine, K., Moss, B. and Balls, H. (1989). The loss of submerged plants with eutrophication .2. Relationships between fish and zooplankton in a set of experimental ponds, and conclusions. *Freshwater Biology*, 22, 89–107.

Jackson, L. J. (2003). Macrophyte-dominated and turbid states of shallow lakes: evidence from Alberta lakes. *Ecosystems*, 6, 213–23.

Jackson, R. D. and Allen-Diaz, B. (2002). State and transition models: response to an ESA symposium. *Bulletin of the Ecological Society of America*, 83, 194–6.

Jeppesen, E., Sondergaard, M., Meerhoff, M., Lauridsen, T. L. and Jensen, J. P. (2007). Shallow lake restoration by nutrient loading reduction—some recent findings and challenges ahead. *Hydrobiologia*, 584, 239–52.

Jones, D. D. (1977). Catastrophe theory applied to ecological systems. *Simulation*, 29, 1–15.

Jones, D. D. and Walters, C. J. (1976). Catastrophe theory and fisheries regulation. *Journal of the Fisheries Research Board of Canada*, 33, 2829–33.

Jones, J. I., Young, J. O., Eaton, J. W. and Moss, B. (2002). The influence of nutrient loading, dissolved inorganic carbon and higher trophic levels on the interaction between submerged plants and periphyton. *Journal of Ecology*, 90, 12–24.

Jones, J. I. and Sayer, C. D. (2003). Does the fish–invertebrate–periphyton cascade precipitate plant loss in shallow lakes? *Ecology*, 84, 2155–67.

Kalm, P. and Gronberger, S. M. (1911). A description of the wild pigeons which visit the southern English colonies in North America, during certain years, in incredible multitudes. *The Auk*, 28, 53–66.

Kelsey, J. and Timmis, J. (2003). Immune inspired somatic contiguous hypermutation for function optimisation. In E. Cantú-Paz, J. A. Foster, K. Deb, L. Davis, R. Roy, U.-M. O'Reilly, H.-G. Beyer, R. K. Standish, G. Kendall, S. W. Wilson, M. Harman, J. Wegener, D. Dasgupta, M. A. Potter, A. C. Schultz, K. A. Dowsland, N. Jonoska and J. F. Miller (eds) *Genetic and evolutionary computation—GECCO 2003, Genetic And Evolutionary Computation Conference, Chicago, IL, USA, July 12–16, 2003. Proceedings, Part I*, pp. 207–18. Springer-Verlag, Berlin.

Kempf, J., Duckstein, L. and Casti, J. (1984). Relaxation oscillations and other non-Michaelian behavior in a slow-fast phytoplankton growth-model. *Ecological Modelling*, 23, 67–90.

Kennelly, S. J. (1987). Inhibition of kelp recruitment by turfing algae and consequences for an Australian kelp community. *Journal of Experimental Marine Biology and Ecology*, 112, 49–60.

Kitzberger, T., Araoz, E., Gowda, J. H., Mermoz, M. and Morales, J. M. (2012). Decreases in fire spread probability with forest age promotes alternative community states, reduced resilience to climate variability and large fire regime shifts. *Ecosystems*, 15, 97–112.

Knowlton, N. (1992). Thresholds and multiple stable states in coral-reef community dynamics. *American Zoologist*, 32, 674–82.

Knowlton, N. (2004). Multiple "stable" states and the conservation of marine ecosystems. *Progress in Oceanography*, 60, 387–96.

Konar, B. and Estes, J. A. (2003). The stability of boundary regions between kelp beds and deforested areas. *Ecology*, 84, 174–85.

van de Koppel, J., Herman, P. M. J., Thoolen, P. and Heip, C. H. R. (2001). Do alternate stable states occur in natural ecosystems? Evidence from a tidal flat. *Ecology*, 82, 3449–61.

van de Koppel, J., Rietkerk, M. and Weissing, F. J. (1997). Catastrophic vegetation shifts and soil degradation in terrestrial grazing systems. *Trends in Ecology and Evolution*, 12, 352–6.

Krasnosel'skii, M. and Pokrovskii, A. (1989). *Systems with hysteresis*. Springer-Verlag, New York.

Kubo, R. (1966). Fluctuation–dissipation theorem. *Reports on Progress in Physics*, 29, 255–84.

Lake, P. S. (2000). Disturbance, patchiness, and diversity in streams. *Journal of the North American Benthological Society*, 19, 573–92.

Langmuir, I. (1989). Pathological science. *Physics Today*, 42, 36–48.

Laycock, W. A. (1991). Stable states and thresholds of range condition on North American rangelands—a viewpoint. *Journal of Range Management*, 44, 427–33.

van Leeuwen, A., De Roos, A. M. and Persson, L. (2008). How cod shapes its world. *Journal of Sea Research*, 60, 89–104.

Lenton, T. M. (2011). Early warning of climate tipping points. *Nature Climate Change*, 1, 201–9.

Leopold, A. (1924). Grass, brush, timber, and fire in southern Arizona. *Journal of Forestry*, 22, 1–10.

Lessios, H. A. (1988). Mass mortality of *Diadema antillarum* in the Caribbean: what have we learned? *Annual Review of Ecology and Systematics*, 19, 371–93.

Lewis, M. A. and Kareiva, P. (1993). Allee dynamics and the spread of invading species. *Theoretical Population Biology*, 43, 141–58.

Lewontin, R. C. (1969). The meaning of stability. In G. M. Woodwell and H. H. Smith (eds) *Diversity and stability in ecological systems*, pp. 12–24. Brookhaven National Laboratory, Upton, NY.

Lilienfeld, S. O. and Landfield, K. (2008). Science and pseudoscience in law enforcement—a user-friendly primer. *Criminal Justice and Behavior*, 35, 1215–30.

Lindenmayer, D. B., Hobbs, R. J., Likens, G. E., Krebs, C. J. and Banks, S. C. (2011). Newly discovered landscape traps produce regime shifts in wet forests. *Proceedings of the National Academy of Sciences of the United States of America*, 108, 15887–91.

Lockwood, J. A. and Lockwood, D. R. (1991). Rangeland grasshopper (Orthoptera, Acrididae) population-dynamics—insights from catastrophe theory. *Environmental Entomology*, 20, 970–80.

Lockwood, J. A. and Lockwood, D. R. (1993). Catastrophe-theory—a unified paradigm for rangeland ecosystem dynamics. *Journal of Range Management*, 46, 282–8.

Loehle, C. (1985). Optimal stocking for semi-desert range—a catastrophe-theory model. *Ecological Modelling*, 27, 285–97.

Loehle, C. (1989a). Forest-level analysis of stability under exploitation—depensation responses and catastrophe theory. *Vegetatio*, 79, 109–15.

Loehle, C. (1989b). Catastrophe-theory in ecology—a critical review and an example of the butterfly catastrophe. *Ecological Modelling*, 49, 125–52.

Loladze, I., Kuang, Y. and Elser, J. J. (2000). Stoichiometry in producer–grazer systems: linking energy flow with element cycling. *Bulletin of Mathematical Biology*, 62, 1137–62.

Loladze, I., Kuang, Y., Elser, J. J. and Fagan, W. F. (2004). Competition and stoichiometry: coexistence of two predators on one prey. *Theoretical Population Biology*, 65, 1–15.

Lotka, A. J. (1956). *Elements of mathematical biology*. Dover Publications, New York.

Louette, G. and De Meester, L. (2007). Predation and priority effects in experimental zooplankton communities. *Oikos*, 116, 419–26.

Loverde-Oliveira, S. M., Huszar, V. L. M., Mazzeo, N. and Scheffer, M. (2009). Hydrology-driven regime shifts in a shallow tropical lake. *Ecosystems*, 12, 807–19.

Lubchenco, J. (1983). *Littorina* and *Fucus*—effects of herbivores, substratum heterogeneity, and plant escapes during succession. *Ecology*, 64, 1116–23.

Lubchenco, J. and Menge, B. A. (1978). Community development and persistence in a low rocky inter-tidal zone. *Ecological Monographs*, 48, 67–94.

MacKenzie, D. I. and Kendall, W. L. (2002). How should detection probability be incorporated into estimates of relative abundance? *Ecology*, 83, 2387–93.

Mantua, N. (2004). Methods for detecting regime shifts in large marine ecosystems: a review with approaches applied to North Pacific data. *Progress in Oceanography*, 60, 165–82.

Mapstone, B. D. (1995). Scalable decision rules for environmental impact studies: effect size, type-I, and type-II errors. *Ecological Applications*, 5, 401–10.

Mashiri, F. E., McClaran, M. P. and Fehmi, J. S. (2008). Short- and long-term vegetation change related to grazing systems, precipitation, and mesquite cover. *Rangeland Ecology and Management*, 61, 368–79.

Mason, N. W. H., Wilson, J. B. and Steel, J. B. (2007). Are alternative stable states more likely in high stress environments? Logic and available evidence do not support Didham et al. 2005. *Oikos*, 116, 353–7.

May, R. M. (1973). *Stability and complexity in model ecosystems.* Princeton University Press, Princeton, NJ.

May, R. M. (1977). Thresholds and breakpoints in ecosystems with a multiplicity of stable states. *Nature,* 269, 471–7.

McCook, L. J. (1999). Macroalgae, nutrients and phase shifts on coral reefs: scientific issues and management consequences for the great barrier reef. *Coral Reefs,* 18, 357–67.

McManus, J. W. and Polsenberg, J. F. (2004). Coral–algal phase shifts on coral reefs: ecological and environmental aspects. *Progress in Oceanography,* 60, 263–79.

Meijer, M. L., de Boois, I., Scheffer, M., Portielje, R. and Hosper, H. (1999). Biomanipulation in shallow lakes in the Netherlands: an evaluation of 18 case studies. *Hydrobiologia,* 408, 13–30.

Menge, B. A. (1976). Organization of a New England rocky intertidal community—role of predation, competition, and environmental heterogeneity. *Ecological Monographs,* 46, 355–93.

Mihaljevic, M., Spoljaric, D., Stevic, F., Cvijanovic, V. and Kutuzovic, B. H. (2010). The influence of extreme floods from the River Danube in 2006 on phytoplankton communities in a floodplain lake: shift to a clear state. *Limnologica,* 40, 260–8.

Milchunas, D. G., Sala, O. E. and Lauenroth, W. K. (1988). A generalized-model of the effects of grazing by large herbivores on grassland community structure. *The American Naturalist,* 132, 87–106.

Möllmann, C., Diekmann, R., Muller-Karulis, B., Kornilovs, G., Plikshs, M. and Axe, P. (2009). Reorganization of a large marine ecosystem due to atmospheric and anthropogenic pressure: a discontinuous regime shift in the central Baltic Sea. *Global Change Biology,* 15, 1377–93.

Morley, F. H. W. (1966). Stability and productivity of pastures. *Proceedings of the New Zealand Society of Animal Production,* 26, 8–21.

Morris, K., Bailey, P. C., Boon, P. I. and Hughes, L. (2003a). Alternative stable states in the aquatic vegetation of shallow urban lakes. II. Catastrophic loss of aquatic plants consequent to nutrient enrichment. *Marine and Freshwater Research,* 54, 201–15.

Morris, K., Boon, P. I., Bailey, P. C. and Hughes, L. (2003b). Alternative stable states in the aquatic vegetation of shallow urban lakes. I. Effects of plant harvesting and low-level nutrient enrichment. *Marine and Freshwater Research,* 54, 185–200.

Morris, K., Harrison, K. A., Bailey, P. C. E. and Boon, P. I. (2004). Domain shifts in the aquatic vegetation of shallow urban lakes: the relative roles of low light and anoxia in the catastrophic loss of the submerged angiosperm *Vallisneria americana. Marine and Freshwater Research,* 55, 749–58.

Muller, E. B., Nisbet, R. M., Kooijman, S., Elser, J. J. and McCauley, E. (2001). Stoichiometric food quality and herbivore dynamics. *Ecology Letters,* 4, 519–29.

Mumby, P. J. (2009). Phase shifts and the stability of macroalgal communities on Caribbean coral reefs. *Coral Reefs*, 28, 761–73.

Mumby, P. J., Hastings, A. and Edwards, H. J. (2007). Thresholds and the resilience of Caribbean coral reefs. *Nature*, 450, 98–101.

Muradian, R. (2001). Ecological thresholds: a survey. *Ecological Economics*, 38, 7–24.

Murdoch, W. W. (1970). Population regulation and population inertia. *Ecology*, 51, 497–502.

Murray, J. D. (1982). Parameter space for Turing instability in reaction diffusion mechanisms: a comparison of models. *Journal of Theoretical Biology*, 98, 143–63.

Murray, J. D. (2002). *Mathematical biology: I. An introduction.* Springer-Verlag, New York.

Murray, J. D. (2003). *Mathematical biology: II. Spatial models and biomedical applications.* Springer-Verlag, New York.

Nagy, K. A. (2005). Field metabolic rate and body size. *Journal of Experimental Biology*, 208, 1621–5.

van Nes, E. H. and Scheffer, M. (2003). Alternative attractors may boost uncertainty and sensitivity in ecological models. *Ecological Modelling*, 159, 117–24.

van Nes, E. H. and Scheffer, M. (2007). Slow recovery from perturbations as a generic indicator of a nearby catastrophic shift. *The American Naturalist*, 169, 738–47.

Ninio, R., Meekan, M., Done, T. and Sweatman, H. (2000). Temporal patterns in coral assemblages on the great barrier reef from local to large spatial scales. *Marine Ecology-Progress Series*, 194, 65–74.

Norström, A. V., Nyström, M., Lokrantz, J. and Folke, C. (2009). Alternative states on coral reefs: beyond coral–macroalgal phase shifts. *Marine Ecology–Progress Series*, 376, 295–306.

Noy-Meir, I. (1975). Stability of grazing systems: an application of predator–prey graphs. *Journal of Ecology*, 63, 459–81.

Noy-Meir, I. (1981). Theoretical dynamics of competitors under predation. *Oecologia*, 50, 277–84.

Odion, D. C., Moritz, M. A. and DellaSala, D. A. (2010). Alternative community states maintained by fire in the Klamath Mountains, USA. *Journal of Ecology*, 98, 96–105.

O'Farrell, I., Izaguirre, I., Chaparro, G., Unrein, F., Sinistro, R., Pizarro, H., Rodriguez, P., Pinto, P. D., Lombardo, R. and Tell, G. (2011). Water level as the main driver of the alternation between a free-floating plant and a phytoplankton dominated state: a long-term study in a floodplain lake. *Aquatic Sciences*, 73, 275–87.

Orians, G. H. (1974). Diversity, stability and maturity in natural ecosystems. In *Unifying concepts in ecology. Proceedings of the First International Ecological Congress*, pp. 64–5. Pudoc, Wageningen.

Orth, R. J., Carruthers, T. J. B., Dennison, W. C., Duarte, C. M., Fourqurean, J. W., Heck, K. L., Hughes, A. R., Kendrick, G. A., Kenworthy, W. J., Olyarnik, S., Short, F. T., Waycott, M. and Williams, S. L. (2006). A global crisis for seagrass ecosystems. *Bioscience*, 56, 987–96.

Osenberg, C. W., Sarnelle, O., Cooper, S. D. and Holt, R. D. (1999). Resolving ecological questions through meta-analysis: goals, metrics, and models. *Ecology*, 80, 1105–17.

Osman, R. W., Munguia, P., Whitlatch, R. B., Zajac, R. N. and Hamilton, J. (2010). Thresholds and multiple community states in marine fouling communities: integrating natural history with management strategies. *Marine Ecology–Progress Series*, 413, 277–89.

Ouimet, C. and Legendre, P. (1988). Practical aspects of modeling ecological phenomena using the cusp catastrophe. *Ecological Modelling*, 42, 265–87.

Pacala, S. W. and Tilman, D. (1994). Limiting similarity in mechanistic and spatial models of plant competition in heterogeneous environments. *The American Naturalist*, 143, 222–57.

Paine, R. T. (1974). Intertidal community structure—experimental studies on the relationship between a dominant competitor and its principal predator. *Oecologia*, 15, 93–120.

Paine, R. T. and Trimble, A. C. (2004). Abrupt community change on a rocky shore—biological mechanisms contributing to the potential formation of an alternative state. *Ecology Letters*, 7, 441–5.

Paine, R. T., Castilla, J. C. and Cancino, J. (1985). Perturbation and recovery patterns of starfish-dominated intertidal assemblages in Chile, New Zealand, and Washington State. *The American Naturalist*, 125, 679–91.

Park, T. (1948). Experimental studies of interspecies competition. 1. Competition between populations of the flour beetles, *Tribolium confusum* Duval and *Tribolium castaneum* Herbst. *Ecological Monographs*, 18, 265–307.

Park, T. (1954). Experimental studies of interspecies competition. 2. Temperature, humidity, and competition in two species of *Tribolium*. *Physiological Zoology*, 27, 177–238.

Persson, L., Amundsen, P. A., De Roos, A. M., Klemetsen, A., Knudsen, R. and Primicerio, R. (2007). Culling prey promotes predator recovery—alternative states in a whole-lake experiment. *Science*, 316, 1743–6.

Peterson, C. H. (1984). Does a rigorous criterion for environmental identity preclude the existence of multiple stable points? *The American Naturalist*, 124, 127–33.

Peterson, C. H. and Bradley, B. P. (1978). Estimating diet of a sluggish predator from field observations. *Journal of the Fisheries Research Board of Canada*, 35, 136–41.

Petraitis, P. S. (1987). Immobilization of the predatory gastropod, *Nucella lapillus*, by its prey, *Mytilus edulis*. *Biological Bulletin*, 172, 307–14.

Petraitis, P. S. (1990). Direct and indirect effects of predation, herbivory and surface rugosity on mussel recruitment. *Oecologia*, 83, 405–13.

Petraitis, P. S. (1998). How can we compare the importance of ecological processes if we never ask, "compared to what?" In W. Resetarits and J. Bernardo (eds) *Experimental ecology, issues and perspectives*, pp. 183–201. Oxford University Press, New York.

Petraitis, P. S. and Dudgeon, S. R. (1999). Experimental evidence for the origin of alternative communities on rocky intertidal shores. *Oikos*, 84, 239–45.

Petraitis, P. S. and Dudgeon, S. R. (2004). Detection of alternative stable states in marine communities. *Journal of Experimental Marine Biology and Ecology*, 300, 343–71.

Petraitis, P. S. and Dudgeon, S. R. (2005). Divergent succession and implications for alternative states on rocky intertidal shores. *Journal of Experimental Marine Biology and Ecology*, 326, 14–26.

Petraitis, P. S. and Hoffman, C. (2010). Multiple stable states and relationship between thresholds in processes and states. *Marine Ecology–Progress Series*, 413, 189–200.

Petraitis, P. S. and Latham, R. E. (1999). The importance of scale in testing the origins of alternative community states. *Ecology*, 80, 429–42.

Petraitis, P. S., Latham, R. E. and Niesenbaum, R. A. (1989). The maintenance of species diversity by disturbance. *Quarterly Review of Biology*, 64, 393–418.

Petraitis, P. S., Methratta, E. T., Rhile, E. C., Vidargas, N. A. and Dudgeon, S. R. (2009). Experimental confirmation of multiple community states in a marine ecosystem. *Oecologia*, 161, 139–48.

Platt, J. R. (1964). Strong inference. *Science*, 146, 347–53.

Portielje, R. and Roijackers, R. M. M. (1995). Primary succession of aquatic macrophytes in experimental ditches in relation to nutrient input. *Aquatic Botany*, 50, 127–40.

Poston, T. and Stewart, I. (1978). *Catastrophe theory and its applications.* Pitman, London.

Potthoff, A. J., Herwig, B. R., Hanson, M. A., Zimmer, K. D., Butler, M. G., Reed, J. R., Parsons, B. G. and Ward, M. C. (2008). Cascading food-web effects of piscivore introductions in shallow lakes. *Journal of Applied Ecology*, 45, 1170–9.

Rabinovich, J. E. (1981). Modelos y catastrofes: enlace entre la teoria ecologica y el manejo de los recursos naturales renovables. *Interciencia*, 6, 12–21.

Rambal, S. (1984). Un modèle de simulations du pâturage en Tunisie présaharienne. *Acta Oecologica–Oecologia Generalis*, 5, 351–64.

Recknagel, F. (1985). Analysis of structural stability of aquatic ecosystems as an aid for ecosystem control. *Ecological Modelling*, 27, 221–34.

Ribbens, E., Silander, J. A. and Pacala, S. W. (1994). Seedling recruitment in forests—calibrating models to predict patterns of tree seedling dispersion. *Ecology*, 75, 1794–806.

Ricklefs, R. E. (1987). Community diversity—relative roles of local and regional processes. *Science*, 235, 167–71.

Rietkerk, M. and van de Koppel, J. (1997). Alternate stable states and threshold effects in semi-arid grazing systems. *Oikos*, 79, 69–76.

Rietkerk, M., Ketner, P., Stroosnijder, L. and Prins, H. H. T. (1996). Sahelian rangeland development; a catastrophe? *Journal of Range Management*, 49, 512–19.

Robinson, A. P., Duursma, R. A. and Marshall, J. D. (2005). A regression-based equivalence test for model validation: shifting the burden of proof. *Tree Physiology*, 25, 903–13.

Rodríguez Iglesias, R. M. and Kothmann, M. M. (1997). Structure and causes of vegetation change in state and transition model applications. *Journal of Range Management*, 50, 399–408.

Roelke, D. L., Zohary, T., Hambright, K. D. and Montoya, J. V. (2007). Alternative states in the phytoplankton of Lake Kinneret, Israel (Sea of Galilee). *Freshwater Biology*, 52, 399–411.

Romano, J. P. (2005). Optimal testing of equivalence hypotheses. *Annals of Statistics*, 33, 1036–47.

Rosenthal, R. (1979). The file drawer problem and tolerance for null results. *Psychological Bulletin*, 86, 638–41.

Rosenzweig, M. L. and MacArthur, R. H. (1963). Graphical representation and stability conditions of predator–prey interactions. *The American Naturalist*, 97, 209–23.

Sale, P. F. (1977). Maintenance of high diversity in coral-reef fish communities. *The American Naturalist*, 111, 337–59.

Sampson, A. W. (1917). Succession as a factor in range management. *Journal of Forestry*, 15, 593–6.

Samuels, C. L. and Drake, J. A. (1997). Divergent perspectives on community convergence. *Trends in Ecology and Evolution*, 12, 427–32.

Sanders, S. R. (1986). *Audubon reader—the best writings of John James Audubon*, Indiana University Press, Bloomington.

Saunders, P. T. (1980). *An introduction to catastrophe theory*. Cambridge University Press, Cambrige, UK.

Savage, M., Sawhill, B. and Askenazi, M. (2000). Community dynamics: what happens when we rerun the tape? *Journal of Theoretical Biology*, 205, 515–26.

Saxon, E. C. and Dudzinski, M. L. (1984). Biological survey and reserve design by Landsat mapped ecoclines—a catastrophe-theory approach. *Australian Journal of Ecology*, 9, 117–24.

Sayer, C. D., Hoare, D. J., Simpson, G. L., Henderson, A. C. G., Liptrot, E. R., Jackson, M. J., Appleby, P. G., Boyle, J. F., Jones, J. I. and Waldock, M. J. (2006). TBT causes regime shift in shallow lakes. *Environmental Science and Technology*, 40, 5269–75.

Sayer, C. D., Burgess, A., Kari, K., Davidson, T. A., Peglar, S., Yang, H. D. and Rose, N. (2010). Long-term dynamics of submerged macrophytes and algae in a small and shallow, eutrophic lake: implications for the stability of macrophyte-dominance. *Freshwater Biology*, 55, 565–83.

Schallenberg, M. and Sorrell, B. (2009). Regime shifts between clear and turbid water in New Zealand lakes: environmental correlates and implications for management and restoration. *New Zealand Journal of Marine and Freshwater Research*, 43, 701–12.

Scheffer, M. (1990). Multiplicity of stable states in fresh-water systems. *Hydrobiologia*, 200, 475–86.

Scheffer, M. (1991). Fish and nutrients interplay determines algal biomass—a minimal model. *Oikos*, 62, 271–82.

Scheffer, M. and Carpenter, S. R. (2003). Catastrophic regime shifts in ecosystems: linking theory to observation. *Trends in Ecology and Evolution*, 18, 648–56.

Scheffer, M. and van Nes, E. H. (2007). Shallow lakes theory revisited: various alternative regimes driven by climate, nutrients, depth and lake size. *Hydrobiologia*, 584, 455–66.

Scheffer, M., Hosper, S. H., Meijer, M. L., Moss, B. and Jeppesen, E. (1993). Alternative equilibria in shallow lakes. *Trends in Ecology and Evolution*, 8, 275–9.

Scheffer, M., Carpenter, S., Foley, J. A., Folke, C. and Walker, B. (2001). Catastrophic shifts in ecosystems. *Nature*, 413, 591–6.

Scheffer, M., Rinaldi, S., Huisman, J. and Weissing, F. J. (2003). Why plankton communities have no equilibrium: solutions to the paradox. *Hydrobiologia*, 491, 9–18.

Scheffer, M., Bascompte, J., Brock, W. A., Brovkin, V., Carpenter, S. R., Dakos, V., Held, H., van Nes, E. H., Rietkerk, M. and Sugihara, G. (2009). Early-warning signals for critical transitions. *Nature*, 461, 53–9.

Schmitz, O. J. (2004). Perturbation and abrupt shift in trophic control of biodiversity and productivity. *Ecology Letters*, 7, 403–9.

Schooler, J. (2011). Unpublished results hide the decline effect. *Nature*, 470, 437.

Schorger, A. W. (1955). *The passenger pigeon: its natural history and extinction*. University of Wisconsin Press, Madison, WI.

Schröder, A., Persson, L. and De Roos, A. M. (2005). Direct experimental evidence for alternative stable states: a review. *Oikos*, 110, 3–19.

Seed, R. (1969). The ecology of *Mytilus edulis* L. (Lamellibranchiata) on exposed shores. I. Breeding and settlement. *Oecologia*, 3, 277–316.

Shurin, J. B., Amarasekare, P., Chase, J. M., Holt, R. D., Hoopes, M. F. and Leibold, M. A. (2004). Alternative stable states and regional community structure. *Journal of Theoretical Biology*, 227, 359–68.

da Silveira Lobo Sternberg, L. (2001). Savanna–forest hysteresis in the tropics. *Global Ecology and Biogeography*, 10, 369–78.

Sinclair, A. R. E. and Fryxell, J. M. (1985). The Sahel of Africa: ecology of a disaster. *Canadian Journal of Zoology–Revue Canadienne De Zoologie*, 63, 987–94.

Skellam, J. G. (1951). Random dispersal in theoretical populations. *Biometrika*, 38, 196–218.

Skrabanek, P. (1987). Skepticism, irrationalism and pseudoscience. *Speculations in Science and Technology*, 10, 191–9.

Skrabanek, P. (1994). The emptiness of the black-box. *Epidemiology*, 5, 553–5.

Slatkin, M. (1974). Competition and regional coexistence. *Ecology*, 55, 128–34.

Slobodkin, L. B. (1961). *Growth and regulation of animal populations*. Holt, Rinehart and Winston, New York.

Sondergaard, M., Liboriussen, L., Pedersen, A. R. and Jeppesen, E. (2008). Lake restoration by fish removal: short- and long-term effects in 36 Danish lakes. *Ecosystems*, 11, 1291–305.

Sousa, W. P. and Connell, J. H. (1985). Further comments on the evidence for multiple stable points in natural communities. *The American Naturalist*, 125, 612–15.

Stansfield, J., Moss, B. and Irvine, K. (1989). The loss of submerged plants with eutrophication. 3. Potential role of organochlorine pesticides—a paleoecological study. *Freshwater Biology*, 22, 109–32.

Stephen, D., Moss, B. and Phillips, G. (1998). The relative importance of top-down and bottom-up control of phytoplankton in a shallow macrophyte-dominated lake. *Freshwater Biology*, 39, 699–713.

Stephens, P. A., Sutherland, W. J. and Freckleton, R. P. (1999). What is the Allee effect? *Oikos*, 87, 185–90.

Stringham, T. K., Krueger, W. C. and Shaver, P. L. (2003). State and transition modeling: an ecological process approach. *Journal of Range Management*, 56, 106–13.

Suding, K. N. and Hobbs, R. J. (2009). Threshold models in restoration and conservation: a developing framework. *Trends in Ecology and Evolution*, 24, 271–9.

Suding, K. N., Gross, K. L. and Houseman, G. R. (2004). Alternative states and positive feedbacks in restoration ecology. *Trends in Ecology and Evolution*, 19, 46–53.

Sutherland, J. P. (1974). Multiple stable points in natural communities. *The American Naturalist*, 108, 859–73.

Sutherland, J. P. (1981). The fouling community at Beaufort, North Carolina—a study in stability. *The American Naturalist*, 118, 499–519.

Sutherland, J. P. (1990). Perturbations, resistance, and alternative views of the existence of multiple stable points in nature. *The American Naturalist*, 136, 270–5.

Thom, R. (1975). *Structural stability and morphogenesis* (transl. D. H. Fowler). Benjamin/Cummings, Reading, MA.

Thomas, J. D. and Daldorph, P. W. G. (1994). The influence of nutrient and organic enrichment on a community dominated by macrophytes and gastropod mollusks in a eutrophic drainage channel—relevance to snail control and conservation. *Journal of Applied Ecology*, 31, 571–88.

Thrush, S. F., Hewitt, J. E., Dayton, P. K., Coco, G., Lohrer, A. M., Norkko, A., Norkko, J. and Chiantore, M. (2009). Forecasting the limits of resilience: integrating empirical research with theory. *Proceedings of the Royal Society B: Biological Sciences*, 276, 3209–17.

Tilman, D., Riech, P. B., Knops, J., Wedin, D., Mielke, T. and Lehman, C. (2001). Diversity and productivity in a long-term grassland experiment. *Science*, 294, 843–5.

Trochine, C., Guerrieri, M., Liboriussen, L., Meerhoff, M., Lauridsen, T. L., Sondergaard, M. and Jeppesen, E. (2011). Filamentous green algae inhibit phytoplankton with enhanced effects when lakes get warmer. *Freshwater Biology*, 56, 541–53.

Trubatch, S. L. and Franco, A. (1974). Canonical procedures for population-dynamics. *Journal of Theoretical Biology*, 48, 299–324.

Turnbull, L., Wainwright, J. and Brazier, R. E. (2008). A conceptual framework for understanding semi-arid land degradation: ecohydrological interactions across multiple-space and time scales. *Ecohydrology*, 1, 23–34.

Underwood, A. J. (1990). Experiments in ecology and management—their logics, functions and interpretations. *Australian Journal of Ecology*, 15, 365–89.

Underwood, A. J. (1991). The logic of ecological experiments: a case history from studies of the distribution of macroalgae on rocky intertidal shores. *Journal of the Marine Biological Association of the United Kingdom*, 71, 841–66.

Underwood, A. J. (2000). Experimental ecology of rocky intertidal habitats: what are we learning? *Journal of Experimental Marine Biology and Ecology*, 250, 51–76.

Valone, T. J., Meyer, M., Brown, J. H. and Chew, R. M. (2002). Timescale of perennial grass recovery in desertified arid grasslands following livestock removal. *Conservation Biology*, 16, 995–1002.

Van Nguyen, V. and Wood, E. F. (1979). Morphology of summer algae dynamics in non-stratified lakes. *Ecological Modelling*, 6, 117–31.

Vandermeer, J., de la Cerda, I. G., Perfecto, I., Boucher, D. and Ruiz, J. (2004). Multiple basins of attraction in a tropicl forest: evidence for nonequilibrium community structure. *Ecology*, 85, 575–9.

Vandermeer, J. H. (1973). Generalized models of two species interactions: a graphical analysis. *Ecology*, 54, 809–18.

Vandermeer, J. H. and Goldberg, D. E. (2003). *Population ecology: first principles.* Princeton University Press, Princeton, NJ.

Viaroli, P., Bartoli, M., Giordani, G., Naldi, M., Orfanidis, S. and Zaldivar, J. M. (2008). Community shifts, alternative stable states, biogeochemical controls and feedbacks in eutrophic coastal lagoons: a brief overview. *Aquatic Conservation–Marine and Freshwater Ecosystems*, 18, S105–S117.

Vitousek, P. (1982). Nutrient cycling and nutrient use efficiency. *The American Naturalist*, 119, 553–72.

Walker, B. H., Ludwig, D., Holling, C. S. and Peterman, R. M. (1981). Stability of semi-arid savanna grazing systems. *Journal of Ecology*, 69, 473–98.

Washington-Allen, R. A., Briske, D. D., Shugart, H. H. and Salo, L. F. (2010). Introduction to special feature on catastrophic thresholds, perspectives, definitions, and applications. *Ecology and Society*, 15(3) [online] <http://www.ecologyandsociety.org/vol15/iss3/art38>.

Webb, S. L. (1986). Potential role of passenger pigeons and other vertebrates in the rapid Holocene migrations of nut trees. *Quaternary Research*, 26, 367–75.

Weiher, E. and Keddy, P. A. (1995). Assembly rules, null models, and trait dispersion: new questions from old patterns. *Oikos*, 74, 159–64.

Westoby, M. (1980). Elements of a theory of vegetation dynamics in arid rangelands. *Israel Journal of Botany*, 28, 169–94.

Westoby, M., Walker, B. and Noy-Meir, I. (1989). Opportunistic management for rangelands not at equilibrium. *Journal of Range Management*, 42, 266–74.

Wilson, J. B. and Agnew, A. D. Q. (1992). Positive-feedback switches in plant communities. *Advances in Ecological Research*, 23, 264–336.

Wissel, C. (1984). A universal law of the characteristic return time near thresholds. *Oecologia*, 65, 101–7.

deYoung, B., Barange, M., Beaugrand, G., Harris, R., Perry, R. I., Scheffer, M. and Werner, F. (2008). Regime shifts in marine ecosystems: detection, prediction and management. *Trends in Ecology and Evolution*, 23, 402–9.

deYoung, B., Harris, R., Alheit, J., Beaugrand, G., Mantua, N. and Shannon, L. (2004). Detecting regime shifts in the ocean: data considerations. *Progress in Oceanography*, 60, 143–64.

Zeeman, E. C. (1976). Catastrophe theory. *Scientific American*, 234, 65–83.

Zimmer, K. D., Hanson, M. A., Herwig, B. R. and Konsti, M. L. (2009). Thresholds and stability of alternative regimes in shallow prairie-parkland lakes of central North America. *Ecosystems*, 12, 843–52.

Index

Note: Page numbers in italics refer to Figures and those in bold refer to Tables.